HOW TO GROW ALMOST EVERYTHING

BY STANLEY SCHULER

Illustrations by Sigman-Ward

M. EVANS AND COMPANY, INC
New York, New York 10017

M. Evans and Company titles are distributed in
the United States by the J. B. Lippincott Company,
East Washington Square, Philadelphia, Pa. 19105;
and in Canada by McClelland & Stewart Ltd.,
25 Hollinger Road, Toronto M4B 3G2, Ontario.

Library of Congress Catalog Card Number 65-13249

ISBN 0-87131-257-3 (Paperback)

ISBN 0-87131-061-9 (Hardcover)

Manufactured in the United States of America

9 8 7 6 5 4 3 2 1

CONTENTS

HOW TO USE THIS BOOK

Few people are of the green-thumb type that can grow anything—even plants they have never seen before.

Most of us need—or at least want—a little guidance. This is particularly true if we have recently moved (as so many of us do these days) to a new part of the country.

Here at last is the information.

How to Grow Almost Everything is the first book ever published that tells how to do exactly that. Unlike other garden encyclopedias, it is not concerned with the fine points of botany. It is, rather, a down-to-the-dirt-under-the-fingernails book that tells you how and when and where to start plants and how to take care of them for the rest of their days.

Whether you are a beginning gardener who is wondering how to grow grass, zinnias and a maple tree, or whether you are an expert uncertain about the culture of an orchid or bromeliad, you will in most cases find the answers here.

Before you dip into the pages that follow, however, please note a few points:

1. *How to Grow Almost Everything* covers the plants growing in the original 48 states. The only plants not included are those which are rarely planted in the home garden because they are indeed rare or have gone out of style (plants do, you know) or are essentially farm crops or grow only in Hawaii or Alaska.

2. *How to Grow Almost Everything* is primarily a book for newer gardeners (although the old hands will find in it a great deal of useful information which is difficult or impossible to locate elsewhere). It does not presume that you own a lot of fancy tools, a coldframe or hotbed or a greenhouse (in fact, nothing at all is said about plants in greenhouses). The ways given for growing plants are the simplest that I know of (for instance, if it is easier to divide a perennial than to raise it from seeds, I suggest only that you divide it).

3. In order to make it as easy as possible for you to find the specific information you need at any given moment, the book is divided into two

main sections. The first is an alphabetical listing of the plants you are likely to want to grow. Here you will find precise directions for raising each plant.

The second section gives the methods for carrying out various basic gardening tasks, such as propagating, seeding, transplanting, improving soil, etc. Brand-new gardeners will undoubtedly find it helpful to read through this section first before they read how to grow a specific plant. Others will probably want to refer to this section from time to time when and if they are uncertain about the ways to handle certain plants.

4. Plants are listed in the How to Grow section under the names by which, in my opinion, they are most widely known. This may be their common name (for example, African violet instead of *saintpaulia*; spruce instead of *picea*) or it may be their botanical name (for example, *gaillardia* instead of gay flower). The only exception to this rule occurs when the entry concerns a genus with a number of species usually identified by their common names. In this case, the entry heading is the botanical name of the genus (for example, *elaeagnus* is the heading under which Russian olive, silverberry and lingaro are mentioned).

If this sounds a bit confusing, don't worry. *How to Grow Almost Everything* has a voluminous index. So if in leafing through the How to Grow section you cannot find a plant under the name that you know it by, just turn to this. It will immediately direct you to the right page.

4. In the interest of saving space (and holding down the cost of the book), I have taken several short-cuts in the How to Grow instructions. Here they are:

If propagating instructions are not given for a plant, it means that for one reason or another you would be well advised to let some one else propagate the plant for you.

Unless the instructions call for a specific type of fertilizer, you can use any of the balanced, commercial fertilizers which are labeled "all purpose" or "for lawns and gardens." (For more about this, see the section on how to improve soil.)

Specific gardening terms are used frequently throughout the book. If you don't understand their meanings, refer to the glossary just before the index.

Stanley Schuler

Greenwich, Conn.

ABELIA Deciduous shrubs to 5′ with clusters of small white or pink flowers in summer. Grows in warm climates; also indoors.

Outdoors. Propagate by stem cuttings in summer. Plant in spring or fall in a sheltered, sunny spot. Provide average, well-drained soil to which considerable extra humus has been added. Water in dry weather. Fertilize in spring. Cut out some of the old stems in winter (flowers come on new wood).

Indoors. Pot in general-purpose potting soil which is kept moist. Grow in a south window in a very cool room. Fertilize several times in late winter, spring and summer.

ACACIA Wattle. Large genus of trees and shrubs to 60′, some thorny, with feathery foliage and handsome yellow flowers in winter and spring. Grows in warmest climates.

Plant young potted plants only. Plant in spring or fall in well-drained, average soil containing a little extra sand and humus. Water well when plants are young, but once they are established they can take care of themselves except in severe drought. (However, *A. riceana* and *A. pruinosa* are two that need regular watering.) Fertilize every year. Prune after flowering.

ACAENA Sheep bur, New Zealand bur. Trailing, spreading evergreen perennials only a few inches high with many small leaflets. Grown in warm or mild climates as a groundcover.

Divide and plant in spring in sun or partial shade in average soil to which some fertilizer has been added. Space 6″ apart. Water regularly while young and growing, thereafter in dry weather. Fertilize in spring if necessary to give a shot in the arm.

ACALYPHA Different species called chenille plant and firedragon plant or copper leaf. Shrubs to 10′, the first with red flower tassels, the other with brilliantly colored foliage. Grows indoors; also in warmest climates.

Indoors. Pot in equal amounts of loam, humus and sand. Grow in a warm south window. Water when soil feels dry. Fertilize only if plant does not seem to be doing well. Repot in fresh soil in the same size pot in spring. At that time, prune top and roots to prevent plant from growing too large.

Outdoors. Propagate in fall by semi-hardwood stem cuttings over heat. Plant in spring in full sun in well-drained loam containing plenty of humus and some sand. Water in dry weather. Fertilize in spring. Prune in spring or summer to control growth. Spray with malathion to control pests.

ACANTHOPANAX Deciduous trees or shrubs, 9′–80′ tall, with very ornamental foliage and clusters of greenish-white flowers followed by black berries. Grows in mild and warm climates.

Plant in spring or fall in sun or light shade in average soil. Water in dry weather. Fertilize every 3–4 years.

ACANTHOSTACHYS STROBILACEA A hanging, epiphytic bromeliad, 2′ long, with orange flowers like a pineapple. Grows indoors, and outdoors in subtropics.

Propagate by cutting off a large offset, hardened somewhat at the base, and planting in potting soil of 1 part sand and 1 part osmunda fiber so that the bottom is just covered. Support until roots form. Water, preferably with rain water, when soil feels dry. Fertilize with a small amount of dilute liquid plant food every six weeks in spring and summer. Grow in an east or west window indoors, or outdoors in partial shade. Give normal house temperature indoors; protect against frost outdoors. Indoors, spray foliage with water in winter and move plant outdoors in summer. Spray with malathion if scale appears; let it stay on plant overnight; then rinse off with clear water.

ACANTHUS MOLLIS Bear's breech. Perennial to 3′ with large notched leaves and white, rose or lavender flower spikes in late summer. Grows in warm climates.

Divide and plant in fall or late winter in partial shade. Grow in average soil in a place where roots, which spread far and wide, can be confined. Water in dry weather and spray foliage with water frequently in summer. Fertilize in spring. Cut back hard in late fall to promote leaf growth. Guard against snails and slugs.

ACHIMENES Trailing rhizomatous plants to 1′ with velvety, gloxinia-like flowers in purple, blue, red and white in summer. Mainly a house plant.

Propagate by division of the rhizomes, by stem or leaf cuttings or by the little cones among the leaves. Pot the rhizomes in January, February or March in 6″ pots filled with sterilized soil composed of equal parts of loam, humus and sand. Lay rhizomes flat and cover with 1″ of soil. Keep in a dark place at about 75° until shoots are well up; then temperature can be reduced somewhat. Grow in an east or west window. Keep soil evenly moist. Fertilize every two weeks. When shoots are 4″ tall, pinch the ends to promote bushing out. After flowering, gradually reduce water until foliage dies. Then cut water off and store the rhizomes in the pot in a dry, cool place. Achimenes may be grown outdoors in the shade in summer. Plant in beds or containers. Handle as above.

ACIDANTHERA Abyssinian wildflower. Plant to 18″ growing from a corm, with large white and purple summer flowers. Grows anywhere but it takes so long to flower that it may be a disappointment in cold climates unless started under glass.

Grow like gladiolus (which see).

ACROCOMIA Palm trees to 40′ with single, bulbous, spindle-shaped trunks and prickly, ragged leaves. Grows in subtropics.

For how to grow, see palm. If the tree is planted in a traveled area, clip off the spines on the trunk.

ACTINIDIA Chinese actinidia, yangtao, tara vine, kolomikta vine, silver vine. Woody vines to 30′ with handsome foliage. Grows in warm climates.

Propagate by half-ripe stem cuttings in summer or by layering. Plant in spring or fall in good, humusy soil in sun or light shade. Don't expose to wind. Water in dry weather. Fertilize in the spring.

Prune in very early spring. Vine needs strong support to twine on.

ADROMISCHUS Different species known as sea shells, pretty pebbles, leopard's spots, plover's eggs. Tiny succulents with thick, prettily colored and shaped leaves. Grows in warmest climates; also indoors.

For how to grow, see cactus, desert.

AECHMEA Large genus of terrestrial bromeliads, to 7′, with tubular or rosette shapes, sometimes colorful leaves with spiny margins, attractive flowers in various colors and sometimes berries. Grows indoors, and outdoors in subtropics.

Outdoors. Propagate by potting the offsets, or suckers, when they are somewhat hard at the base, very shallowly in a mixture of one part sand and one part shredded osmunda. Grow on in this. Keep cups of rosette-shaped plants filled with water, preferably rain water. Water other types when soil feels dry. Fertilize with a little dilute liquid plant food every six weeks in spring and summer. Grow in light shade. Protect from frost. Spray with malathion if scale appears, and rinse the plant the following day with water.

Indoors. Grow as above in an east or west window. Spray foliage with water every few days. Move outdoors in summer.

AESCHYNANTHUS Lipstick vine. Sometimes identified as *trichosporum.* Small shrubs trailing several feet when grown indoors, with red or yellow flowers from spring through fall. A house plant.

Propagate by stem cuttings. Pot in equal parts of sphagnum moss and osmunda with some charcoal chunks mixed in. Grow in a south window, but protect from noonday sun in summer. When growing and flowering, water regularly, feed with liquid fertilizer monthly, spray foliage with water frequently and give a temperature of 70° or better. When resting in winter, withhold fertilizer, water sparingly and give a lower temperature. Don't spray foliage. Prune to 6″–12″ in the fall.

AEONIUM Shrubby succulents to 3′ with rosettes of fleshy leaves on woody stems and large flower clusters in various colors in late winter or early spring. Grows in warmest climates; also indoors.

For how to grow, see cactus, desert. Plants die after flowering.

AFRICAN TULIP TREE *Spathodea campanulata.* Evergreen tree to 40′ with showy scarlet flowers in winter and spring. Grows in subtropics.

Plant in fall in sun in average, well-drained soil. Water in dry spells, especially in winter and spring. Fertilize every couple of years.

AFRICAN VIOLET *Saintpaulia.* Evergreen house plant, about 6″, with blue, purple, white, pink, red and mixed-color flowers almost continuously.

Propagate by leaf cuttings in water or moist sand, or divide plants that have two or more crowns. Plant in a sterilized mixture of equal parts of loam, humus and sand. Be sure pot is scrubbed clean. Use soap. Grow in a warm east or west window. Ventilate frequently but don't expose plants to cold drafts. Keep away from icy cold windowpanes.

Water thoroughly when soil surface feels nearly dry. You can stand pot in a container of water or water from above; but in the latter case, use tepid water and if you get any on the leaves, let them dry before exposing them to sun. To provide necessary humidity, stand pots on wet pebbles and occasionally spray foliage with tepid water.

Fertilize every two weeks with any plant food that contains twice as much phosphorus as nitrogen and potassium. Follow manufacturer's directions carefully. African violets need considerable feeding, but don't overdo it.

Repot plants only once a year unless the roots crowd the pots. Use fresh sterilized soil.

If plants are attacked by cyclamen mites, which cause the center leaves to become brittle, shiny and twisted, set pots in dilute sodium selenate solution until soil is soaked through. To prevent petiole, or leaf-stem, rot on plants in red clay pots, cover rims of pots with melted paraffin or aluminum foil.

AGAPANTHUS Lily of the Nile, African lily. Vigorous, tuberous-rooted, mainly evergreen plants to 4′ with many clusters of blue or white flowers. If grown in a tub (which is the best procedure), they can be grown almost everywhere.

Propagate by division of roots, but don't expect flowers for several years. Plant in March or April in a tub filled with general-purpose potting soil with double quantity of humus. Crowns should be just at the surface. Keep soil evenly moist, and after growth starts feed every two weeks with liquid fertilizer. Grow in a south window. Move outdoors into sun when weather is safely warm. Continue watering and fertilizing until flowering stops. Then gradually withhold water and bring plant indoors into a cool, dry place. Water only about once a month—just enough to keep tubers from shriveling—until you start plant into growth again.

In subtropical climates, agapanthus can be planted directly in the garden but does better in a tub—especially when rootbound. It is left outdoors all year.

AGAVE Succulents to 40′ with leaves arranged in rosettes. One of the best-known species is the century plant. Grows in hot, arid climates; also indoors.

For how to grow, see cactus, desert. Grow pot plants outdoors for as many months as your climate allows.

AGERATUM Floss flower. Annual to 1′ with dense, fluffy flowers mostly blue. Grows anywhere. Sometimes grown as a house plant.

Outdoors. Sow seeds indoors about eight weeks before last frost. Grow in sun or light shade in average soil. Fertilize once after plants are established. Keep watered.

Indoors. In late summer, cut back small garden plants and pot in general-purpose potting soil. Keep moist and in light shade. Growth will soon start again. Move indoors before frost and grow in a cool, sunny window.

AILANTHUS ALTISSIMA Tree of heaven, stinkweed. Smooth-barked deciduous tree to 60′ that grows well in cities in temperate climates.

Plant female trees only in early spring or fall. Grow in sun in average, well-drained soil. Water in dry weather. The tree takes care of itself. It is nothing to brag about, even at best.

AJUGA Bugle, carpenter's weed. Perennials, one a popular groundcover, with blue or white spring flowers, to 10″. Grows in all but coldest climates.

Divide plants in spring or sow seeds any time up to July for flowering plants the next year. Grow in sun or shade in average soil which is improved in spring with a little fertilizer. Water regularly. If plants are killed by crown rot in humid weather, remove at once and sterilize the soil.

AKEBIA QUINATA Five-leaf akebia. Woody vine to 20′ used for shade, screening and as a groundcover. Grows in all but very cold climates. Is evergreen in mild climates.

Propagate by stem cuttings or root divisions, or sow seeds in spring. Grow in sun or very light shade in fertile, well-drained soil. Water in dry weather. Cut back hard in early spring or fall; if you don't, plant will take over your place. Feed in spring only if it needs a shot in the arm for some reason. Provide strong support to grow on.

ALDER Alnus. Deciduous trees or shrubs to 70′ with elliptical leaves. Grows in mild climates.

Plant in spring or fall in wet or moist average soil in the sun and forget it.

ALLAMANDA One species known as golden trumpet (*A. cathartica hendersoni*); another as yellow bell or bush allamanda (*A. neriifolia*). Evergreen climbers to about 30′ or shrubs to 6′ with lovely flowers that are usually yellow but in one species purple. Grows outdoors in subtropics. Golden trumpet is also grown indoors.

Outdoors. Propagate in spring by stem cuttings. Plant in spring in good soil which should be enriched with fertilizer every 2–4 weeks from spring to fall while plant is growing and blooming its best. All species require full sun except yellow bell, which also does well in partial shade. Water regularly during growing season. Prune after blooming. Tie vines to a sturdy support. If you live in a frost area, pot plants in the fall and bring indoors or into a protected place.

Indoors. Pot in general-purpose potting soil. Keep moist and feed every month. Spray foliage with water frequently. Grow in a south window. Control mealybugs with malathion. You had better move plant outdoors in summer and cut back hard after it has bloomed.

ALLIUM The onion and its relatives. Plants called *alliums* grow from bulbs to 3′ and produce globular, often large clusters of small purple, white, pink or yellow flowers. Grows almost everywhere.

Plant bulb in fall or early spring in average, well-drained soil in sun. Cover with soil to three times the bulb's own depth. Water regularly. Fertilize in spring. Propagate by the bulbils in the fall.

ALOCASIA Plants growing from rhizomes, to 10′ but usually smaller, with large, beautiful, often colored leaves. A house plant.

Propagate in late winter by suckers which include a piece of the rhizome. Pot in equal parts of loam, humus and sand. Provide thick drainage layer in bottom of pot. Grow in an east window in a warm room away from drafts. Keep soil moist. Fertilize every 4–6 weeks while plant is making growth from March on. Spray leaves with water every day during this period. Move outdoors in summer into semi-shade if you wish. Reduce water and humidity in fall and winter.

ALMOND *Amygdalus communis.* Deciduous tree to 20′ with pink or white spring flowers and peach-like fruits containing a nut. Nut production is limited to warm climates but trees grow further north.

Plant two different varieties, such as Nonpareil and Ne Plus Ultra, to assure nuts. Plant in early spring in a sunny location in well-drained, average soil. Space 30′ apart. Water deeply in dry weather. Apply well-rotted manure in fall. Keep soil cultivated.

Pruning consists of cutting back tree, when planted, to about 30″. At the end of the first year, prune to three main branches equally spaced around the trunk and about 8″ apart up and down. Thereafter, prune very little. However, after tree has been bearing five years, prune annually—removing some of the smaller branches (fruit is borne on the same wood for about five years).

ALOE Succulents of many shapes and sizes (up to 60′) with soft, fleshy leaves arranged in rosettes and red or yellow flowers. Grows in warmest climates; also indoors.

For how to grow, see cactus, desert. The aloes are very easy to grow and undemanding as long as the soil is well-drained and they receive full sun. They can stand low temperatures but not freezing. To prevent cuttings from rotting, press cut end on a hot iron before planting.

ALPINIA Often called ginger. Plants growing from rhizomes to 10′ with fine foliage and very showy, multicolored flowers in summer. Grows outdoors in warmest climates and can be grown indoors.

Divide rhizomes in spring. Each one must have at least one bud. Plant 2″ deep in the equivalent of two parts loam, two parts humus and one part sand. Add a good supply of well-rotted manure if available. Grow in a warm spot in filtered sun or light shade. Keep soil moist. Fertilize every month while plants are making growth. Reduce water somewhat and do not feed after flowering. Do not expect flowers the first year.

Indoors, alpinia is grown in the same way but also requires constant spraying with water.

ALSTROEMERIA Peruvian lily, parrot lily. Plants to 4′ with clusters of tubular, yellow or rose flowers in summer. Grows almost everywhere.

In spring, when weather is reliably warm, plant the roots 4″–6″ deep in rich, humusy, well-drained soil in partial shade. Fertilize when growth appears. Water in dry spells. In the fall, when foliage dies down, dig up roots and store them in slightly moist sand in a cool, dark place. In the deep south, however, plants can stay in the ground all year.

ALTERNANTHERA Joseph's coat. Perennials to 30″ (but smaller indoors) with pretty red or variegated foliage. Grows indoors; also outdoors in warmest climates.

Indoors. Propagate by young stem cuttings. Pot in general-purpose soil. Grow in a south window. Keep moist. Fertilize lightly with dilute liquid plant food several times during the year. Move outdoors in sun or partial shade in summer. Plant directly in the garden if you wish. Foliage can be sheared hard during summer.

Outdoors. In the south, plant in light shade in average, well-drained soil. Water in dry weather. Fertilize in late winter or early spring. Keep pruned, otherwise plant will become leggy.

ALYSSUM Madwort. Perennials to 15″, mostly with yellow flowers in spring or summer. Most popular species is known as basket of gold or gold dust. Grows almost everywhere.

Propagate by division in spring or fall or by seeds in summer. Grow in the sun in average soil which needs water only in dry spells. Apply a little fertilizer in early spring.

AMARANTHUS Different species called love-lies-bleeding, tassel flower, prince's feather, Joseph's coat, fountain plant, summer poinsettia. Annuals to 3′ with red foliage. Grows almost everywhere.

Sow seed in sunny location in the garden after last frost. Only average soil is required. Keep watered. Fertilize when plants are thinned to 12″ with a general-purpose plant food rich in nitrogen.

AMARYLLIS *Hippeastrum.* Bulbous plant to 3′ with flowers like enormous lilies in reds, pinks, white and multi-colors. A house plant; also grows outdoors in warmest climates.

Indoors. Plant bulb in general-purpose potting soil in late fall or winter. Use a pot only 2″ wider than the bulb. Allow half of the bulb to extend above the soil surface. Insert a stout stake. Water and set in a warm, not too light place until growth starts. Then move into a warm south window and water when soil begins to feel dry. When flowers fade, cut off stalk 2″ above the bulb. Do not cut or damage the leaves. Continue watering regularly and apply liquid plant food every month. When weather turns warm, you can follow one of two courses of action either keep plant indoors and continue watering and feeding, or plunge the pot in partial shade in the garden (place a thick layer of gravel or hard cinders under it), protect plant against animals, and continue watering and feeding. In the fall, before frost, bring pot indoors. Then, whether pot was outdoors or in during the summer, gradually withhold water and let the leaves yellow. About the end of October, cut them from the bulb and store the bulb in the pot in a cool, dark place. Water only occasionally to keep bulb from shriveling. Then about January, bring pot into a warm, not too light room and start into growth again.

Outdoors. Buy bulbs and plant in fall or winter in a sunny location in average, neutral, well-drained soil. In warmest climates (southern Florida, for example) the necks of the bulbs are left exposed; elsewhere, they are covered with a little soil. Keep watered when bulbs are making growth. After flowering, until September, fertilize monthly. Spray garden with malathion or DDT to keep down insects which may attack amaryllis or, more likely, carry disease. If leaves of a plant become streaked or spotted, lift bulb and soak it (but not the green growth) overnight in 1 tsp. Lysol to 1 gal. water. Amaryllis bulbs produce offsets which can be broken off and raised to blooming size, but it takes several years.

AMAZON LILY *Eucharis grandiflora.* Bulbous plant to 1′ with fragrant white flowers several times a year. Grows indoors; also outdoors in warmest climates.

Indoors. Pot in the spring in equal parts of loam, coarse humus and sand. The pots must have a thick drainage layer. Necks of bulbs should just show above the surface. Water sparingly until growth starts, then keep soil evenly moist. Grow in an east or west window except during winter, when plant should be in a south window. Keep warm. To get repeat bloom after the first flowering, let the soil dry out until the plant begins to look sick; then water thoroughly. After 4–6 weeks of this, start watering regularly again and fertilize lightly once a month.

Outdoors. Plant in light shade in rich, well-drained, humusy soil enriched with a little bone meal. Just cover the bulbs and space them 1′ apart. Keep well watered. Fertilize every other month when new growth appears in early spring. Cover plants if frost threatens. Divide bulbs after they have flowered but only every few years when they become crowded: they do not like to be disturbed.

AMPELOPSIS Monkshood vine, pepper vine. Vigorous vines to 30′ with large leaves and clusters of berries that change color. Most of the popular species can be grown almost everywhere.

Ampelopsis is not particular about soil. Propagate by stem cuttings or by layering, and plant in spring in sun or partial shade. Water in dry weather. Fertilize in spring. Spray with malathion to control caterpillars. If trained upwards, provide a sturdy support with wires or other members which the tendrils can twine around. Plant can also be allowed to run across the ground. Cut back hard in early spring.

ANACAHUITA *Cordia boissieri,* Texas wild olive. Evergreen tree to 15′ with delightful white flowers in spring and summer. Grows in warmest climates.

Plant in spring or fall in full sun or light shade. Grow in average, well-drained soil. Apply lime if soil is acid. Needs little attention otherwise.

ANCHUSA Alkanet, bugloss, summer forget-me-not. Perennials to 5′ with brilliant blue flowers. One species a biennial usually treated as an annual. Grows in all but warmest climates.

Perennial. Propagate by seeds planted in spring, or by division of roots in spring or fall. Plant in average, well-drained soil in full sun. Fertilize in early spring and again in summer. Keep soil evenly moist. Stake tall plants. Plants may flower a second time if cut back after first flowering. Mulch lightly in winter in severe climates, but be careful not to cover crowns, because they may rot.

Annual. Sow seeds indoors eight weeks before last frost. Handle as above when moved outdoors.

ANEMONE Windflower. Perennials to 30″ with large, very showy flowers in many colors. Poppy anemone (*A. coronaria*) grows in warm climates, but other tuberous types are hardy further north. Japanese anemone (*A. japonica*) grows in all but warmest climates.

Poppy anemone and other tuberous species. Plant tubers in fall in rich, humusy, well-drained soil in partial sun. They should not be more than 2″ deep, 6″ apart, and with fuzzy side up. Water sparingly until growth appears, then increase supply. Fertilize two or three times during the growing season. Mulch in winter in colder parts of the south.

Japanese anemone. Divide plants in spring (but not too often because anemones do not like to be disturbed). Plant in rich, humusy, well-drained soil in early spring in shade. Water regularly. Fertilize twice during spring and summer. Spray with malathion in the summer to control blister beetles. Mulch in winter.

ANGELICA ARCHANGELICA Perennial herb to 5′ grown for its seeds and stems. Grows in cool climates.

Sow seeds in late summer or fall and move the plants in spring into sun or partial shade in average soil containing some humus. Water regularly. Fertilize when plants are established.

ANGELONIA ANGUSTIFOLIA Perennial 1′ tall with violet, snapdragon-like flowers. Grows in warmest climates.

Propagate by young stem cuttings in spring. Plant in fall in sun or partial shade in average soil. Water in dry weather. Fertilize a couple of times in winter and spring.

ANISE *Pimpinella anisum.* Annual herb to 2′, grown for the leaves and seeds. Grows almost everywhere.

Sow seeds outdoors in sun when ground is warm. Thin to 9″. Water in dry weather. Fertilize when thinned.

ANTHEMIS Chamomile, golden marguerite. Perennials to 3′ with daisy-like yellow flowers. Grows in all but most extreme climates.

Divide plants or sow seeds in spring. Grow in average soil in a sunny spot. Water only in very dry weather. Fertilize every other year in spring unless you want plants to spread faster.

APACHE PLUME *Fallugia paradoxa.* Deciduous shrub to 3′ with white flowers and fruits with feathery tails. Grows in southwest.

Plant in spring or fall in sun in average, well-drained soil. Water until established.

APHELANDRA Tropical evergreen shrubs to 4′ with fine foliage and large yellow or red flower spikes. Grows indoors.

Pot in general-purpose potting soil containing some extra humus. Grow in a warm east or west window. Keep soil moist except after flowering, when water supply should be reduced for several months. Fertilize every other month. To prevent plant from becoming too leggy too fast, pinch out new growth. Start new plants from stem cuttings every year.

APOSTLE PLANT *Neomarica,* twelve apostles, false flag. Two-foot plant growing from a rhizome with 12 sword-like leaves arranged in an iris-like fan and with fragrant white-and-violet flowers. A house plant.

Pot in general-purpose potting soil

in a container that is only a little larger than the rhizome. Cover with about ½″ of soil. Grow in an east or west window until bloom is about to start, then shift to a south window. Keep warm and water faithfully. Spray foliage often with water. Fertilize when growth starts and once or twice thereafter. Divide rhizome after flowering.

APPLE *Malus pumila*. Familiar fruit which grows almost everywhere but best where temperature changes are not excessive. Trees range up to about 50′ but dwarf and semi-dwarf varieties are available.

Plant two varieties together in full sun in a high location. Space trees 40′ apart. For most varieties the soil should be a deep, rich, well-drained loam containing a substantial amount of humus. Fertilize every spring at the rate of 1 lb. per year of age up to a maximum of 15 lb. This should be scattered on the ground around the tree and scratched and watered in. Water deeply in dry spells.

Prune trees in early spring. Cut back one-year-old whips when you plant them to a single bud about 4′ above the ground. Next spring, select three branches spaced evenly around the trunk and about 1′ apart up and down, and remove all others. In the following springs, remove all branches except the previously selected three and others needed for a well-shaped tree. Pruning of bearing trees consists of light thinning out of small branches. To stop a tree from growing too tall (20′ is a desirable height), let the leader grow until a good lateral branch develops at the desired height; then cut off the leader just above this branch.

In June, after some of the fruits drop, thin out others that are undesirable so that there is only one apple per spur.

To protect young trees against mice and rabbits, encircle the trunks lightly with ¼″ wire mesh. Sink this in the ground and let it extend at least 2′ up the trunk.

To control insects and disease, spray as follows: In late winter, when buds swell, apply a dormant oil spray. From the time the buds begin to open until four weeks after petals fall, apply all-purpose fruit spray every week. Thereafter, until two weeks before harvest, apply all-purpose spray every fortnight.

APRICOT *Prunus armeniaca*. Ornamental deciduous tree to 25′ with pink spring flowers and small peach-like fruits. Grows in cool and warm climates.

Plant one-year-old trees in spring. The trees need full sun but should be planted on a north slope or in some spot where they do not receive so much sun in late winter that they bloom in advance of last frost. The soil should be deep, well-drained and of above-average quality. Space trees 25′ apart. Water deeply every two or three weeks during dry season. Fertilize in early spring at the rate of 1 lb. per year of age up to a maximum of 10 lb.

Head back young trees at planting time to 30″ and remove side branches. The next year, remove all but the leader and two branches which are on opposite sides of the tree and at slightly different heights. The next year, head back the leader and two main branches slightly and remove excess branches growing from the main trunk. Continue in this way to develop a wide spreading but stubby tree. When fruiting starts, remember that one-year shoots bear some fruit but that spurs on the branches carry much more. The spurs produce for only three years. Therefore, you should cut out bearing branches when they are four years old.

To control insects and diseases, follow schedule for spraying peaches (which see).

AQUATIC PLANTS See also water lily and lotus. Following are some of the many perennials which grow under water: *Elodea canadensis,* also called anacharis, ditch moss, water pest or water thyme; *Cabomba,* also called fanwort, water shield, Washington grass; *Ludwigia* or swamp loosestrife; *Hippuris vulgaris* or mare's tail; *Myriophyllum* or water milfoil; *Sagittaria*; *Vallisneria spiralis*, also called channel grass, eel grass or tape grass. Following are plants — mainly annuals — which grow in water and partly float on the surface: *Nymphoides peltatum* or floating heart (perennial); *Hydrocharis morsus-ranae* or frog's bit; *Salvinia; Pistia stratiotes,* also called shell flower or water lettuce; *Trapa natans* or water chestnut; *Ceratopteris thalictroides* or water fern; *Limnocharis humboldti* or water poppy. All these plants grow almost everywhere.

Propagate the perennials either by division of the roots or, where applicable, by layering of runners. Buy new annual plants. Plant in spring when the weather is warm in a container of ordinary soil. Keep pot submerged in a pool exposed to the sun. Trim plants as necessary to control growth. In warm climates the perennials will go through the winter underwater. Elsewhere, start each year with new plants.

ARABIS Rock cress, wall cress, Perennials to 1′ with fragrant white flowers in spring. Grows best in cool climates.

Sow seeds up to midsummer, or divide plants after flowering. Grow in sun in average, sandy, well-drained soil. Feed a little bone meal after flowering, and prune plants slightly at the same time. Water only in long dry spells. Mulch in winter.

ARALIA Different species known as Hercules' club or devil's walking stick; spikenard; elk clover; angelica tree; udo; bristly sarsaparilla or wild elder; wild sarsaparilla. A mixed-up genus of perennials, shrubs and trees to 25′, the latter two usually spiny. Grows mainly in warm climates, but some species grow well north. See also *dizygotheca.*

Plant in the spring, the perennials in light shade, the others in sun or light shade. Most species prefer a rich, moist, heavy soil, but wild sarsaparilla and Chinese angelica tree also grow in dry, sandy soil. Divide perennials in the spring, if necessary, and fertilize every spring. Prune shrub species grown as hedges in spring, and cut out all dead or weak wood.

ARBORVITAE *Thuja.* One species, *T. occidentalis*, known as white cedar. Evergreens of various shapes to 60′ with fan-like sprays of scale-like leaves. Grows best in cool, moist climates, especially in the northeast and northwest. See also *Hiba arborvitae.*

Do not grow arborvitae in a smoky, industrial atmosphere. Plant in spring or early fall in a sunny location protected to some extent from the wind. Grow in average, well-drained, humusy soil. Water regularly in dry weather. Prune in spring. Fertilize old, sparsely foliaged plants in spring. Spray with malathion if infested with red spider.

ARBUTUS One species known as strawberry tree; another as madrona, Pacific madrone, laurelwood or Oregon laurel. Broadleaved evergreen trees to 80′ with white or pink flowers and red fruits. Grows in warm climates. See also trailing arbutus.

Plant in spring or fall in sun in average, well-drained soil with considerable humus. Needs protection from wind. Fertilize every two to three years. Water in dry spells. Prune after flowering to shape plant. Spray the strawberry tree with malathion to control aphids.

ARCHONTOPHOENIX One species called king palm. Palm trees to 100′ with slender, ringed trunks usually enlarged at the base and with spreading fans of leaves. Grows in warmest climates.

For how to grow, see palm.

ARCTOTIS African daisy. Annual to 2′ with daisy-like flowers in various colors. Grows anywhere.

Sow seeds outdoors where plants are to grow when ground is warm. For somewhat earlier bloom, start seeds indoors. Give full sun, average soil. Thin plants to 10″–24″, depending on height of the variety. Fertilize when thinning is done. Water in dry weather.

ARDISIA CRISPA Coralberry. Tropical shrub to 3′ with shiny foliage and clusters of red berries in fall. Grows outdoors in warmest climates; also indoors.

Outdoors. Propagate by seeds in the spring or semi-ripe stem cuttings. Plant in spring in light shade in rich, humusy soil. Water in dry weather. Fertilize when growth starts and once or twice thereafter until fruit is set. Prune in early spring to control growth and shapeliness.

Indoors. Pot in equal parts of loam, humus and sand. Grow in a north window at 60°–65°. Water when soil feels dry. Fertilize lightly every two months while making growth. Move outdoors into shade in summer.

ARECASTRUM One species called queen palm. Palm trees to 40′ with solitary, ringed trunks and loose top clusters of fans. Grows in subtropics.

For how to grow, see palm. Plants start slowly but then develop rapidly. They do not live very long.

ARENGA One species called black sugar palm. Palm trees to 40′ with trunks often covered with black sheaths, and with blunt leaves and flowers which bloom from the top down, and red and yellow fruits. Subtropical.

For how to grow, see palm.

ARGEMONE Different species called Mexican poppy or crested poppy. Annuals to 3′ with prickly leaves and stems and orange, yellow, white or purple flowers. Grows almost everywhere.

Sow seeds where plants are to grow after danger of frost is past. Grow in sun in average soil. Thin plants to 2′ apart and fertilize when this is done. Water in dry weather.

ARMERIA Thrift, sea pink. Often listed as *statice.* Tufted evergreen perennials to 1′ with rounded, white or rose flowers in late spring. Grows in cool climates.

Divide in spring or fall and plant in full sun. The plants grow in average soil but prefer one that is sandy. Water only in periods of severe drought. Fertilize very lightly in early spring. Divide plants every 2–3 years.

ARROWHEAD *Sagittaria.* Erect marsh plants to 5′ with arrow-shaped leaves. Different species are found in different regions.

Divide roots in early spring and plant in half shade in rich, wet soil or even in soil under a few inches of water.

ARTEMISIA Wormwood. Different species called southernwood, mountain sage, mugwort, old woman, old man, dusty miller, absinthe plant. See also tarragon. Large genus of perennials and small shrubs to 4′ with handsome aromatic foliage. Grows in cool and warm climates.

Divide and plant in spring. Grow in full sun in well-drained, sandy, not too fertile soil. Do not feed and do not water (except *A. albula* and *A. vulgaris* need somewhat more moisture than others).

ARTICHOKE *Cynara scolymus,* globe artichoke. Perennial to 5′ producing more or less circular, green flower buds with succulent leaves and stems. Best grown in warm climates where ground does not freeze. Also see Jerusalem artichoke.

Artichokes require good, well-drained soil which is deeply dug and mixed with humus and fertilizer. Sow seeds indoors in February and transplant seedlings to individual pots. Keep in warm, sunny window. Set plants outdoors when danger of frost is past. Plants should be in full sun in an out-of-the-way spot. Space 3′ apart in rows 4′–5′ apart. Keep weeded and watered. Apply liquid fertilizer twice during the summer or sidedress with dry fertilizer. Harvest buds while they are still compact. In fall, burn tops when they die down. Mulch over winter with humus and wood ashes.

ARUNCUS Goat's beard. Often erroneously called *spiraea.* Four-foot perennial with white flower plumes in June. Grows in temperate climates.

Grow like *astilbe* (which see).

ARUNDINARIA Genus of bamboos. Grows in warmest climates.

For how to grow, see bamboo.

ASH *Fraxinus.* Deciduous tree to 120′, prized for its wood. One has large white flowers. Different species to be found from the Rockies to the east coast and in the far west. See also mountain ash.

Plant in early spring or fall in a sunny situation. Most ashes need soil of only average quality; however, the blue ash does best when lime is added. Black and pumpkin ash require a moist soil. All trees benefit from fertilizer every 2–4 years.

ASPARAGUS *Asparagus officinalis.* The vegetable. Grows almost everywhere.

Grow in full sun in average, well-drained soil with a pH of 6·5 or higher. Mix in well-rotted manure, or humusy material and general-purpose fertilizer. Early in the spring, dig trenches 8″–10″ deep and 4′ apart. Set in one-year-old roots of a rust-resistant variety such as Mary Washington. Space plants 18″ apart in the row. Spread roots horizontally and cover with 2″–3″ of soil. As stalks grow, gradually add soil until it is level with the ground. Keep well watered. Cultivate regularly to keep out weeds, but be careful not to work soil too deeply lest you disturb the roots.

Asparagus is not ready to be harvested for two years. Then the stalks are cut just below the soil level with a sharp knife. The cutting period lasts for about eight weeks. Then let the plants grow. In the fall, before seeds have set, cut off stalks 6″ above ground. No winter protection is necessary as a rule.

Feed plants heavily early every spring with general-purpose fertilizer. Use well-rotted manure if available.

ASPARAGUS FERN *Asparagus plumosus and A. sprengeri* (also called emerald feather). Vines to 6′ with feathery, fern-like foliage. Usually grown indoors, but sometimes outdoors in subtropics.

Indoors. Soak seed overnight in water; then plant in general-purpose potting soil. Keep damp and in the dark until growth starts. Then grow in a cool east or west window. Keep well watered. Spray foliage with water frequently. Fertilize lightly when making growth. Train *A. plumosus* up strings. *A. sprengeri* can be handled the same way or allowed to trail.

Outdoors. Grow in semi-shade. Soil should be moist but well drained; rich and humusy. Keep well watered. Train branches to a light trellis. Prune only as needed and at any time. Spray with malathion to control insects.

ASPIDISTRA Cast-iron plant. Evergreen plant to 3′ with large, leathery shiny leaves. Grows indoors; also outdoors in subtropics.

This is a tough plant and does well even when neglected. But better results come with somewhat better care. Divide in early spring and plant in well-drained, general-purpose potting soil or the equivalent outdoors. Water regularly. Spray foliage with water occasionally. Grow in any light. Fertilize sparingly once or twice a year (do not, however, fertilize varieties with variegated foliage, because they may lose their variegation).

ASPLENIUM Spleenwort. One popular species (*A. nidus*) called bird's nest fern—a house plant with undivided 3′ fronds in a bird's-nest-like clump. Other species have fine, feather fronds up to 2′. These grow outdoors in temperate climates.

For how to grow, see fern.

ASTER The handsome annual aster, to 30″, which blooms in reds, white, blues and purple in midsummer and fall is the China aster. It does best in not-too-hot climates. Perennial asters, to 5′, which bloom in late summer and fall are of many types, grow almost everywhere.

Annual. Sow seeds outdoors either in a seedbed or where plants are to grow after danger of frost is past. For earlier bloom, start seeds indoors about six weeks before they can go outside. For a long season of bloom, make several sowings of seeds at two-week intervals.

Annual asters grow in full sun or partial shade. Soil should be well drained, deeply dug, enriched with humus. Fertilize at monthly intervals. Water regularly. Cultivate frequently but not too deeply. To control the various ailments that attack asters, do not plant them in the same location year after year. Burn plants in the fall. Spray or dust every ten days with Sevin and zineb. Plant only wilt-resistant types.

Perennial. Asters may be started from seeds sown outdoors in spring, but it is better to start with purchased plants or to make divisions in early spring. Grow in average soil in the sun. Fertilize in spring and again before flowering starts. Water in dry weather.

ASTILBE False spirea, goat's beard, meadowsweet. Perennials to 3′ with dense spikes of white, pink or purple flowers. Grows in temperate climates.

Divide and plant in early spring. *Astilbe* prefers partial shade but grows in sun. Soil should be well drained, rich, humusy. Keep moist. Mulch in summer to hold in moisture and in winter to protect against bitter weather. Fertilize in early spring and once more about two months later. Spray with malathion to control red spider.

AUBRIETIA False rock cress. Mat-forming perennial to 6″ with red to purple flowers in spring. Grows in warm climates.

Divide plants or sow seeds in spring. Grow in sun in light, well-drained soil. Water in dry weather. Fertilize lightly in early spring. Shear foliage a little after flowering.

AUSTRALIAN BLUEBELL CREEPER *Sollya heterophylla.* A delicate, evergreen, trailing plant to 6′ with clusters of blue flowers in summer. Grows in warm climates.

Propagate by stem cuttings in the spring. Plant in moist, humusy soil. Fertilize in spring. Grow in sun or light shade. Prune in early spring. Spray with malathion to control aphids.

AUSTRALIAN FLAME TREE *Brachychiton acerifolium.* Almost evergreen tree to 60′ with red summer flowers and large black fruits. Grows in subtropics.

Plant in spring or fall in partial shade in average soil. Water in dry spells. Fertilize every 2–3 years.

AUSTRALIAN SILK OAK *Grevillea robusta.* Evergreen tree to 100′ with feathery foliage and orange flower clusters. Small specimens are grown indoors. Tree also grows outdoors in warmest climates.

Outdoors. Plant in early spring or fall in sun in average, well-drained soil. Water well until established; thereafter, you should not have to worry about water except in severe drought. Fertilize every 4–5 years. Cut out broken limbs.

Indoors. Sow seeds indoors in winter or early spring. Transplant into 6″ pots filled with general-purpose potting soil. Keep moist. Grow in a cool (under 60°) east or west window. Fertilize every couple of months. Don't expect flowers.

AVOCADO *Persea americana.* Evergreen tree to 50′, producing large green or black, edible fruits. Grows in subtropics; also indoors.

Outdoors. Plant in spring in a high, sunny spot that is not exposed to strong wind. The soil should be extremely well drained (the roots cannot stand long in water), of average quality with a little extra humus added. To insure fruiting, it is necessary to plant (35′ apart) at least two different but compatible varieties of avocado; for example, Pollock and Waldin, or Taylor and Eagle Rock, or Lula and Fuerte.

During spring, summer and fall of the first two years apply general-purpose fertilizer every two months at the rate of about ½ lb. per tree. Thereafter, fertilize in late winter, late spring and fall at the rate of 2 lb. per year of age of the tree up to a maximum of 10 lb.

Water deeply in dry weather, especially when fruit begins to set. Prune after fruiting to control size and shape. Spray with a miticide to control red spider.

Indoors. Avocados are easily propagated by potting the seeds, large end down, in general-purpose potting soil. Cover with ½″ of soil. Or set a seed in a narrow tumbler of water so that its base just touches the water; and pot in soil when good roots develop. Once rooted, keep the plant in a warm, sunny window. Water when soil feels dry. Fertilize every two months. Move outdoors in summer.

AZALEA *Rhododendron.* Deciduous and evergreen shrubs under 15′ with gorgeous, many-colored flowers in spring. Grows in all but the bitterest dry climates if you get the right types. Also a house plant.

Outdoors. Plant in spring in the north, in fall in the south. Grow in sun or partial shade in a spot protected from winter winds. Soil should be deeply dug, well drained, neither too heavy nor too light, acid. Mix in an ample quantity of peatmoss. To increase acidity, mix in aluminum sulfate. Set plant at the same depth that it previously grew. Water thoroughly at planting time; thereafter, water heavily in dry weather and in the fall before the ground freezes. Fertilize annually in early spring and again about the middle of summer with Hollytone. Do not cultivate soil under azaleas but keep it covered at all times with a mulch of oak leaves, pine needles or peatmoss (which is less good because it is not so acid). Remove dead flowers. Prune, if necessary, by cutting out dead wood and pinching tips of branches to promote bushiness. Spray with malathion if insects appear. Propagate by stem cuttings in summer, but note that many species do not root easily.

Indoors. Gift azaleas do nicely indoors for a short time, but to grow azaleas indoors year after year is not easy. Keep plants delivered from florist in a sunny window in a very cool (50°–60°) room and keep soil evenly moist, but not soggy, at all times. In the spring, hardy azaleas can be planted outdoors in the garden in mild climates and left there. Or keep the plant in its pot and plunge the pot in the garden in light shade. Keep watered and spray foliage with water frequently. Cut back long branches to maintain shape. In the fall, before frost, bring plant indoors and store at 50° or less in a light but not sunny place. Water only when soil is about to dry out completely. In January, apply fertilizer, start watering regularly and move the plant into a cool, sunny window to bloom.

AZALEAMUM Trick name given various chrysanthemum hybrids. See chrysanthemum.

AZARA Evergreen shrub to 18′ with small glossy leaves, inconspicuous, vanilla-scented flowers and orange berries. Can be trained as a vine or espaliered. Grows in warm California.

Plant in spring or fall in shade in average soil. Keep watered. Fertilize after flowering. Pinch stem ends and prune lightly after flowering.

BABIANA Cormous plants to 10″ with a fan of iris-like leaves and purple or red flowers in spring. Grows outdoors in warmest climates; also indoors.

Outdoors. Plant in fall in sun in average, well-drained soil. Corms should be set 2″ deep and spaced 3″ apart. Water in dry weather. Fertilize in winter. Dig up every 2–3 years to separate the small corms.

Indoors. Grow like *freesia* (which see).

BABY'S BREATH *Gypsophila,* mist, gypsum pink, chalk plant, fairybreath, gauzeflower. Annuals and perennials to 3′ with clouds of white or pink flowers. Grows almost everywhere.

Sow seeds in spring in average soil in sun or light shade. The soil must be well drained, and if it is acid it must be limed. Keep watered and cultivated. Fertilize in early spring. Mulch perennials in winter. Do not disturb established plants.

BABY'S TEARS *Helxine solieroli,* Irish moss, Corsican carpet plant. Small, moss-like creeper with minute leaves. A house plant; also grows outdoors in warm climates.

Divide plant at any time and plant in general-purpose potting soil or the equivalent outdoors. Keep soil moist at all times. Grow in a north window or outdoors in shade.

BACCHARIS Different species called groundsel bush, chaparral broom, mule fat. Deciduous and evergreen shrubs to 10′, mainly useful for holding drifting sand and soil. Different species grow in the east and on the west coast.

Dig from the wild or propagate by stem cuttings. Plant in spring or fall in sun in almost any well-drained soil. Water until plants are established.

BALD CYPRESS *Taxodium distichum,* southern cypress, pond cypress. Coniferous trees to 150′ with small needles which are shed in the fall, a trunk that bulges at the base and "knees" growing up from the roots. Grows mainly in the south.

Plant in spring or fall in sun in average soil that is, preferably, moist or even wet. Feeding every 3–4 years stimulates growth, but it takes many years for large trees to develop in any case.

BALL CACTUS *Notocactus.* Round desert cactus to 3′ with brightly colored spines. Different species called sun cup, Indian head. Grows in warmest climates; also indoors.

For how to grow, see cactus, desert.

BALLOON FLOWER *Platycodon.* Perennial to 30″ with balloon-shaped buds that open into white or blue, bell-shaped flowers. Grows in all but extreme climates.

Divide in spring and plant in well-drained, average soil in the sun. Fertilize in early spring. Water in dry spells. Cultivate lightly and not deeply. Stake plants, otherwise they will topple.

BALSAM *Impatiens balsamina.* Annual to 3′ with flowers in various pastel colors close to the stem. Grows almost everywhere; also indoors.

Outdoors. Start seeds indoors about six weeks before last frost, or sow outdoors where plants are to grow when all danger of frost is past. Grow in full sun. Soil should be well dug, rich, somewhat humusy. Fertilize when young plants are well established and again 4–6 weeks later. Keep well watered.

Indoors. Grow like patience (which see).

BALSAM Balsam fir, *Abies balsamea.* See fir.

BAMBOO Various families of grasses with straight, hollow, woody stems up to 50′. Some types form compact clumps; others spread widely by means of running roots. Grows in temperate to warmest climates. See also Mexican bamboo.

Divide plants in spring or early fall and plant in sun or light shade in average soil that is well drained. Plant running types in containers or restricted garden pockets unless you want a forest. Water well until established; then water only in dry weather. Fertilize in spring, especially if plants are container grown. Cut excess canes at the base when young; remove old canes in winter.

BAMBUSA Large genus of bamboos. Grows in warmest climates.

For how to grow, see bamboo.

BANANA *Musa cavendishi.* This dwarf (about 6′) banana grows fairly well outdoors in southern Florida but is more reliable—though it rarely produces fruit—indoors.

Plant suckers or rhizomes 1′ deep in general-purpose potting soil with a double quantity of humus and also some well-rotted or dried manure. Grow in full sun in a warm room. Keep soil moist at all times. Fertilize every 2–3 weeks. Move outdoors in summer.

BANANA SHRUB *Michelia fuscata.* Fifteen-foot evergreen, magnolia-like shrub with cream-edged-with-red flowers with a banana fragrance. Grows in warmest climates.

Plant in spring or fall in average soil in partial shade. Water in dry weather. Fertilize every couple of years. Prune in winter.

BANEBERRY *Actaea,* cohosh. Perennial wildflower to 2′ with white flower clusters and red or white berries in fall. Grows mainly in cool and cold climates, although the white baneberry is also found in the south.

Divide roots in early spring or late fall. Plant in partial shade in humusy woods soil that is slightly acid and holds moisture.

BARBERRY *Berberis.* Large genus of deciduous and evergreen shrubs to 8′ with small leaves which are brightly colored in fall, yellow flowers and red or black fruits. Grows almost everywhere except coldest climates.

Propagate by seeds planted outdoors in the fall or by cuttings of young wood in June. Plant in spring or fall in sun or light shade. All species do well in average, well-drained soil which retains some moisture; but deciduous types also grow in quite dry soil. Prune in late winter to keep plants within reasonable size and to control shape. Fertilize in spring if plants need a boost.

Note that some species are an alternate host for wheat rust and should not be planted in wheat-growing regions.

BARREL CACTUS Several genera of desert cacti, to 10′, more or less barrel-shaped. The spines are often hooked. Grows indoors; also outdoors in hot climates.

For how to grow, see cactus, desert.

BASIL *Ocimum.* Annual to 18″ grown for the leaves, which are used in flavoring. There is also an ornamental purple-leaved variety. Grows almost everywhere.

Sow seeds in the sun in average soil when it is warm. Thin plants to 1′ and fertilize at that time. Water in dry weather.

BASKET FLOWER *Centaurea americana.* Annual to 4′ with large ray flowers, related to the cornflower. Grows best in warm climates.

Grow like cornflower (which see). Plant in spring only, however.

BASKET GRASS *Oplismenus compositus.* Sometimes identified as *O. hirtellus.* Sprawling 4″ grass with striped leaves. Grows indoors.

Divide plants and pot in general-purpose potting soil. Grow in a warm east or west window. Keep soil moist, and to provide humidity, stand pots on wet pebbles.

BAUHINIA Different species known as orchid tree or mountain ebony; St. Thomas tree; butterfly flower or Jerusalem date. Trees under 30′, shrubs and vines with showy, many-colored flowers and strange foliage. Subtropical.

Propagate by seeds sown in pots. Plant in permanent position when specimen is still small and is dormant. Bauhinias need full sun and average soil that is well drained. Little special watering is required except in unusual dry spells. Fertilize twice a year. Cut back vine types 50 per cent after flowering. Train these to a sturdy support.

BEACH PLUM *Prunus maritima.* Deciduous shrub to 6′ with white flower clusters and purple-black, edible fruits. Grows in coastal areas in the northeast.

Plant in spring or fall in sun in average, sandy, well-drained soil near the ocean. Water until established. Fertilize in the spring for the first couple of years. Thereafter, like most native plants, the shrub should pretty well take care of itself.

BEAN (GREEN, SNAP or STRING) *Phaseolus vulgaris.* Familiar green or yellow beans which are usually eaten in their entirety. Bush beans grow to about 21″ tall; pole beans to 7′. Grows almost everywhere. See also lima bean.

Bush beans. Plant in full sun in average soil into which general-purpose fertilizer has been mixed. Sow seeds 1″ deep and 2″–3″ apart in rows 18″ apart after all danger of frost has passed. Make succession sowings at 2–3 week intervals. Cultivate regularly to keep down weeds, but only when foliage is dry. Water in dry weather. Fertilize when plants are up several inches if you wish. Dust plants regularly with rotenone or pyrethrum as soon as the Mexican bean beetle larvae appear on underside of leaves.

Pole beans. Grow as above. Plant in rows along a fence or trellis on which vines can grow. Or plant 4–6 beans in a cluster around a sturdy 7′ pole. Space poles 3′ apart. Fertilize pole beans once or twice as they grow.

BEAN (SHELL) *Phaseolus.* Bush beans to about 24″, producing large pods containing fat, various colored beans which are usually eaten after they have been dried. Grows almost everywhere.

Red kidney, white marrowfat, navy beans. Grow like bush beans (see entry above). But allow pods to mature before harvesting. Then cut down the bushes and allow pods to dry.

Fava, or broad, bean. Grow like bush beans (see entry above). This variety, however, must be planted as soon as the ground can be worked in the spring. It does not tolerate heat.

BEARBERRY *Arctostaphylos uva-ursi,* kinnikinnick. Spreading evergreen plant only a few inches high with small white flowers and red berries. Grows in mild and cold climates.

Start with potted plants. Plant in early spring in light shade. The soil should be very well supplied with sand and humus. Water regularly until plants are established; thereafter, they require little attention.

BEAUTY BUSH *Kolkwitzia amabilis.* Deciduous shrub to 10′ with grey-green foliage and bell-shaped, pink flowers with yellow throats in late spring. Grows best in mild climates.

Plant in early spring or fall in sun in average, well-drained soil. Water in dry weather. Fertilize every 2–3 years in early spring. Prune rather severely after flowering.

BEEBALM *Monarda,* bergamot, Oswega tea, red balm, horse mint. Perennials to 3′ with striking flowers and bracts in various colors. Grows almost everywhere.

Divide in spring and plant in partial shade or sun. The soil should be well drained, of average quality with extra humus added. Fertilize in early spring and again about June. Keep soil moist, especially if plants are in full sun. Divide about every three years.

BEECH *Fagus.* Mostly very large deciduous trees to 100′ with grey bark and dense foliage which may be dark green, copper or purple, depending on the variety. Grows mainly east of the Mississippi.

Plant in early spring or fall in an open space where the tree will have plenty of room to spread (it is too big for the average small yard). Grow in well-drained, light, rich loam. Be careful in transplanting not to damage the taproot, although if the specimen has been well grown in a nursery, this should not present a problem. Loosen soil which becomes badly compacted under the trees. Fertilize every 3–5 years. In New England, spray in winter with oil emulsion to control scale insects and a serious disease. Do not try to grow beeches in cities.

BEET *Beta vulgaris.* Familiar vegetable to about 20″ grown mainly for its red, bulbous root but also for its succulent young leaves. Grows best in cool climates.

Grow in full sun in average (but not stony) soil that has been cultivated deeply and enriched with fertilizer. Sow seed $\frac{1}{2}$″ deep in rows 18″ apart as soon as soil can be worked in the spring. Make succession sowings. Thin plants to 2″–4″ when leaves are about 4″ tall.

When plants are about 6″ high, sprinkle nitrate of soda between the rows to hasten growth. Cultivate carefully. Water in dry weather.

BEGONIA Bewildering family of perennials with red, pink, white, yellow, or orange flowers. They are either fibrous rooted or grow from tubers and rhizomes. Some are low, some are tall, some are trailing. Most types can be grown indoors. Some are also grown outdoors but are usually treated as annuals. See also tuberous begonia.

Wax begonias (B. semperflorens) indoors. Propagate by tip cuttings taken from short stems in spring and started in damp sand or a glass of water. Pot rooted plants in equal parts of sterilized loam, humus and sand. Don't pack soil too tight. Water when soil feels dry (don't use very hard or very soft water if you can avoid it). Fertilize every 2–3 weeks until flowering stops. Stand pots on wet pebbles in a warm east or west window. Spray foliage with water a couple of times a week. Pinch stem ends to promote bushy growth. Keep dead leaves and flowers picked off. If an old plant becomes ratty looking, cut old stems to the ground when flowering stops or when spring growth begins. Move plants outdoors into partial shade in summer and plunge in the ground (except in the south, where pots should be above ground). Repot in the next size larger pot in fresh soil in the fall. Spray with malathion to control cyclamen mites or thrips, which cause deformed buds and flowers.

Wax begonias outdoors. Propagate by tip cuttings or, if you want a mass of plants, sow seeds indoors 3–4 months before last frost. Plant outdoors in average, humusy soil in partial shade. Keep moist. Fertilize monthly. Spray foliage with water in dry weather. Pinch stem ends to promote bushiness. Plants can be treated as perennials in warmest climates; elsewhere, take cuttings in late summer to make house plants.

Calla-lily begonias indoors. Propagate by stem cuttings which have leaves that are at least half green (if the leaves are more than half white, the cutting won't root). Grow like wax begonias but at 60°–65°. Don't move plants around the room; they react badly.

Angel-wing begonias indoors. Propagate by stem cuttings or division of plants. Grow like wax begonias. Pinch stem ends regularly to control growth, and keep old stems cut out after flowering.

Angel-wing begonias outdoors. Handle like wax begonias outdoors. Angel-wings can stay outdoors all year round in warmest climates.

Hairy-leaved begonias indoors. Handle like wax begonias.

Maple-leaf begonias indoors. Handle like wax begonias. Take stem cuttings in the fall to make sure you have a continuous supply of plants. Old ones do not always survive the winter.

Basket begonias indoors. Handle like wax begonias. Propagate by layering or cuttings. Cut back hard after flowering or in spring before new growth starts. Keep stem ends pinched.

Rhizomatous begonias, such as the beefsteak begonia, indoors. Grow like wax begonias but with the following attention: Propagate by leaf cuttings taken with 2″ stems and inserted in moist sand at an angle. Plant rhizomes near one side of the pot by placing them atop (not in or under) the soil. The growing tip of the rhizome should be aimed at the opposite side of the pot so that it has lots of soil to crawl across. Grow in a south window in winter. Be sure to water only when soil surface feels completely dry; and water very sparingly in fall and early winter.

Rex begonias indoors. Grow like wax begonias but with the following attention: Propagate by leaves taken with 2″ stems. Insert stems in a glass of water or moist sand and cover the leaves with polyethylene film to hold in moisture. When roots develop and new growth starts, pot up in soil with the new growth resting on the soil surface. Cut off old leaf when little plant is well established.

Rex begonias use the same soil as wax begonias but need a little extra humus. Keep soil moist but not soggy. Grow in a warm east window or a warm, very bright north window. Spray foliage daily with water. Pinch off flower buds if you wish.

Rex begonias outdoors. Grow these only in warmest climates in shade and out of the wind. Handle as above.

Christmas begonias indoors. The winter-flowering tuberous begonias are difficult to grow. Just get the most you can out of florists' plants by keeping them in an east window at a daytime temperature of no more than 70° and a night-time temperature of 60°. Stand pots on wet pebbles. Keep soil moist. Don't wet leaves. When plants finally stop blooming, throw them out.

Tuberous begonias. See tuberous begonia.

BELLS OF IRELAND *Molucella laevis,* shellflower. Three-foot, scented annual with green bells grouped around the stems. Grows almost everywhere.

Sow seeds indoors 8–10 weeks before last frost. Move outdoors into a sunny location in average soil. Water regularly. Fertilize when plants are established.

BENZOIN AESTIVALE Spice bush, Benjamin bush. Deciduous shrub to 15′ with aromatic foliage, yellow flowers in early spring. Widely distributed in the east.

Propagate by layering. Plant in spring or fall in average, moisture-retentive soil in light shade. Water frequently. Fertilize every couple of years.

BERGENIA Densely foliaged perennials to 18″ with clusters of pink flowers in spring. Grows in temperate climates.

Grow like *saxifrage* (which see).

BIGNONIA CAPREOLATA Cross vine, trumpet flower. Beautiful, dense, evergreen vine to 40′ with clusters of trumpet-shaped, orange-red flowers in late spring and early summer. Clings to rough surfaces by tendrils. Grows in warm climates.

Propagate by stem cuttings and plant in well-drained, fertile soil in sun or partial shade in the spring. Use as a groundcover or to cover walls. In the latter case, provide something for the vine to climb on. Keep weak branches thinned out. In the spring, cut back as much as 10′ if necessary. Fertilize at that time. Water in dry spells.

BILLBERGIA One popular species known as queen's tears. Mainly epiphytic bromeliads, to about 3′, of various shapes and with incredibly colored flowers often coupled with bright pink or red bracts. Grows indoors or outdoors in subtropics.

Grow like *aechmea* (which see). Give some sun.

BIRCH *Betula.* The best species are the white, or canoe, birch, to 100′, which grows in cold, moist climates, and the European birch, to 60′ with not quite as white bark, which grows equally far north and further south. The shape of the latter varies; some varieties are about as pendulous as the weeping willow.

Plant in early spring or fall in a sunny location in average, well-drained but moist soil which is slightly acid. Fertilize every 2–3 years. Water in very dry weather. Remove dead branches as they occur and dress all wounds with tree paint. Spray with malathion in spring and perhaps again about July 1 to control leaf miners. Or apply a systemic poison. Cut out dark sunken areas (cankers) in bark.

BIRD OF PARADISE *Strelitzia reginae.* Four-foot perennial with spectacular blue and orange flowers resembling a bird in flight. Grows in warmest climates; also indoors.

Outdoors. Divide plants in early spring and plant in full sun in a soil consisting of equal parts of loam, humus and sand with bone meal added. The crown of the plant should be level with the soil surface. Water regularly when growing, but reduce supply in winter. Fertilize from spring through fall at 4–6 week intervals. Don't count on plant's blooming until it has at least seven leaves.

Indoors. Handle as above. Plant in a large pot or tub. Keep at 55°–65° in winter. Move outdoors in sun in summer.

BITTER ROOT *Lewisia rediviva.* Perennial only a few inches tall with leaves in a rosette and wheel-like white flowers that last a day. Native to the Rocky Mountains but grows in any cool or cold climate which does not have too much rain in winter.

Transplant from the wild in early fall. Grow in full sun in sandy, deep, well-drained soil containing some humus. Needs little attention. Don't be surprised if leaves die after plant flowers. Growth will start again in the fall.

BITTERSWEET *Celastrus.* Rugged, rampant, deciduous vine to 25′ with spectacular orange-yellow fruit clusters in fall. Grows almost everywhere.

Propagate by stem cuttings or by layering. Plant several bittersweet vines fairly close together to assure fruiting. Bittersweet grows in average soil in full sun or partial shade. It is not bothered by wind, cold or lack of moisture Do not fertilize too much, because it grows fast anyway. Provide a very strong support for it to clamber on, and steer the stems where you want them to go. Cut back in early spring every year; if you don't, the vine will run wild. Spray with malathion if attacked by scale insects.

BLACKBERRY *Rubus.* Prickly plant to 6′ or more with seedy black fruits. Grows in almost every area where there is enough summer moisture.

Propagate from suckers. Plant in early spring in a sunny location in average, well-drained but moisture-retentive soil. Space plants 3′ apart in rows 6′ apart. The plants should be set a little deeper than they formerly grew. Cut the tops back to

12″–18″. Fertilize with general-purpose plant food early every spring. Mulch between rows with leaves, grass clippings, etc. This will keep down weeds, eliminate need for cultivation (which must be done very carefully—if it is done—because of spreading, shallow roots) and provide humus for the soil the next year. Water regularly up to September 1; thereafter, only enough to protect plants from fall drought. Pull out (don't just cut off) suckers between rows but keep some of those that develop in the rows. If a plant becomes covered with an orange rust, remove and burn it at once; spray remaining plants with ferbam.

Pruning is done as follows: After fruit has been picked, cut out canes which bore it. Young, non-bearing canes should be cut back to 18″ when they grow somewhat taller than this; then, early the next spring, lateral branches which have developed should be cut back to 18″.

BLACK CALLA *Arum palaestinum*, Solomon's lily. Tuberous plant to 15″ with blackish, calla-lily-like flowers. Grows indoors.

Grow like *caladium* (which see).

BLADDER FERN *Cystopteris*. Ferns to 3′ with long, slender fronds. Grows in temperate climates.

For how to grow, see fern.

BLADDER NUT *Staphylea*. Deciduous shrubs or trees to 15′ with clusters of white flowers and membranous seed capsules. Grows in cool climates.

Plant in spring or fall in partial shade in moist, average soil. Water in dry weather. Fertilize in spring.

BLEEDING HEART *Dicentra spectabilis*. Perennial to 2′ with sprays of deep pink, heart-shaped flowers in spring. Grows in all but extreme climates.

Divide plants in early spring. Plant in partial shade in well-drained, humusy soil. Keep moist in spring and summer. Fertilize in spring. Do not disturb or divide until plants get too large for the border.

BLETILLA A bulbous terrestrial orchid, to 15″, which produces dainty, little, purple flowers resembling the florist's orchids. A house plant; also grows outdoors in temperate and warm climates.

Indoors. Pot bulbs in general-purpose potting soil. Set just below the surface. Grow in an east or west window. Keep moist. Fertilize every three weeks. Keep foliage growing after flowering stops; then when spring weather warms, plunge the pot in a lightly shaded spot in the garden. Continue watering and fertilizing. In fall, let foliage die down; stop watering and fertilizing, and store pot in a cool, dark place for two months. Then start into growth again.

Outdoors. Plant in spring in partial shade in average, well-drained soil. Water and fertilize regularly. Divide clumps, when they grow too large, in spring.

BLOOD LEAF *Iresine herbsti*, chicken gizzard. Six-foot shrub with purple-red leaves. A house plant; also grows outdoors in warmest climate.

Indoors. Propagate by cuttings of young stems in late winter. Pot in general-purpose potting soil and grow in a warm south window. Water regularly. Fertilize every 6–8 weeks. Move outdoors into sun in summer. Plants may also be taken from their pots and set directly in the garden.

Outdoors. In the north, house plants or cuttings from house plants can be planted directly in the garden for the summer. In the deep south, they can be left in the ground the year round. Fertilize about three times in spring and summer. Prune as hard as necessary to control growth.

BLOOD LILY *Haemanthus.* Bulbous plants to 20″ with large, round clusters of flowers, usually red. Grows outdoors in warmest climates; also indoors.

Outdoors. Plant bulbs 1′ apart in well-drained, average soil in partial shade in early spring. Soil should just cover the tops of the bulbs. Water regularly when growth starts. Feed once or twice. After foliage dies down in the fall, do not water until growth starts again.

Indoors. Pot in general-purpose potting soil. The pot should be only 2″ wider than the bulb. Upper third of bulb should be exposed. Water freely when plant is growing and give a little liquid plant food every 3–4 weeks. After flowering, gradually reduce water and stop feeding. When foliage dies, stop watering altogether and store bulb in its pot in a dark, cool place.

BLOODROOT *Sanguinaria canadensis*, redroot, Indian plant. Perennial wildflower growing from a red root, with large irregular leaves and waxy, white flowers on 8″ stems. Grows in cool climates east of the Rockies.

Divide roots in late summer or fall and plant in partial shade in rich, humusy, slightly acid soil that holds moisture. Plant dies down by middle of summer.

BLUEBERRY *Vaccinium.* Handsome deciduous shrubs to 15′, grown for their small, edible, blue berries. Highbush varieties grow where there is some freezing weather but temperature does not fall below −20°. Rabbiteye varieties grow in warm climates.

Plant at least two different varieties of blueberries 4′–5′ apart in rows 8′–10′ apart. Plant in a sunny location in early spring. The planting hole should be 24″–30″ deep and have in the bottom a thick layer of coarse drainage material. The soil must contain a large amount of peatmoss or other acid humus, and sand, and have a pH of 5.0 or less. To acidify soil, mix in aluminum sulfate. Set the plants 1″ deeper than they formerly grew. Water well for the first several weeks; thereafter, once a week in dry spells.

Apply a couple of handfuls of general-purpose fertilizer early every spring, starting the second year. To reduce the need for cultivation and maintain soil acidity, keep a light mulch of oak leaves or pine needles around the plants. Prune the bushes annually in early spring after the third year. Cut out low-hanging branches as well as some, but not all, of the main stems which are more than three years old. Also remove the small side shoots that have borne fruit, because they will not bear again. Fruit comes only on shoots that grew the previous year.

Rabbiteye varieties are handled in the same way except that they are spaced 15′ apart. They are usually planted in midwinter.

BLUE LACE FLOWER *Trachymene coerulea.* Also identified as *didiscus.* Annual to 30″ with blue flowers like those of Queen Anne's lace. Grows best in cool climates.

Sow seed indoors eight weeks before last frost, or outdoors when weather is warm. Grow in full sun in average soil that is fertilized several times with general-purpose plant food. Water in dry weather. Pinch plants to promote bushiness.

BLUE SAGE *Eranthemum nervosum,* blue crossandra. Tropical evergreen shrub to 5′ with lovely blue flowers in spikes. Grows indoors; also outdoors in subtropics.

Propagate by stem cuttings in spring after flowering. Plant in general-purpose potting soil (or the equivalent outdoors) and grow in partial shade. Good drainage is essential. Keep moist. Fertilize about every six weeks in late fall and winter. Cut back hard after flowering stops in spring. Pinch stem ends frequently to make plant fill out.

BLUE SPIREA *Caryopteris incana,* bluebeard. Deciduous shrubs to 5′ with clusters of blue flowers in late summer and fall. Grows best in warm climates.

Propagate by stem cuttings in summer. Plant in spring or fall in full sun in average soil containing sand. Needs little water except in dry weather. Feed lightly in spring. Cut back hard in fall. If killed to the ground by frost, the plant should send up new growth which will bloom the following fall.

BOLTONIA Perennials to 8′ with airy foliage and white or purple, aster-like flowers. Grows in all but extreme climates.

Divide in spring or fall and plant in average soil in sun. Water in dry weather. Fertilize in spring.

BORAGE *Borago.* Annual herb to 2′ with small, bright-blue flowers. Grows almost everywhere.

Sow seeds in rows 18″ apart in average soil early in the spring. Thin plants to 1′ and fertilize at that time. Give sun. Water in driest weather. Spray with nicotine sulfate to control aphids.

BOSTON FERN *Nephrolepis exaltata bostoniensis.* Favorite of all house ferns, with narrow, more or less open, ultimately drooping fronds to 5′ long.

For how to grow, see fern. The Boston fern is very tolerant of house conditions but needs plenty of moisture at the roots and prefers a temperature of about 65° or a little less. Move outdoors into shade in summer. Propagate by pinning the runners down to the soil and covering with a little soil.

BOSTON IVY *Parthenocissus tricuspidata.* Wall-clinging, deciduous vine to 35′, related to Virginia creeper. Grows in all but coldest climates.

Grow like Virginia creeper (which see).

BOTTLE BRUSH *Callistemon.* Evergreen shrubs to 30′ with red flower spikes resembling bottle brushes. Grows outdoors in warmest climates; also indoors.

Outdoors. Plant in spring or fall in sun in average, well-drained soil. The bottle brushes do not mind wind or salt spray. Little water is required except in the case of *C. viminalis*. Fertilize every 1–2 years in early spring. Prune fairly severely in early fall to remove dead wood and prevent legginess.

Indoors. Pot in general-purpose potting soil. Grow in a south window at 65° or under. Water when soil feels dry. Fertilize every two months. Move outdoors in summer. Prune lightly in late summer.

BOUGAINVILLEA Paper flower. Evergreen vine to 20′ with mainly red flowers. Grows in subtropics; also indoors.

Outdoors. Propagate by semi-hardwood stem cuttings in spring. Plant in full sun in spring in no better than average, well-drained soil (*bougainvillea*, despite its lush beauty, does not do well in rich soil). Water in dry weather. Fertilize in late winter and again in the fall with a general-purpose fertilizer with a very low nitrogen content. Vines bloom best in winter but some flowers may appear at other times. After heavy winter flowering, cut plant back one-third to one-half. Tie stems to strong support. Spray with malathion to control caterpillars.

Indoors. Pot in general-purpose potting soil and grow in a south window. Keep moist. Feed and prune as above. Repot only when vine no longer grows well. Move outdoors in summer.

BOWIEA VOLUBILIS Climbing onion. Sometimes erroneously identified as *schizobasopsis volubilis*. A bulbous succulent that puts out twining 7′ stems that bear small, greenish-white flowers. Grows indoors; also outdoors in warmest climates.

Plant bulbs in the fall in soil used for cacti (see cactus, desert). Leave the top two-thirds of the bulb exposed. Fertilize lightly. Grow in full sun in a warm place. Water when soil feels dry. Gradually reduce water in the spring and store plant in a dry place in summer.

BOX *Buxus*, boxwood. Handsome, dense, evergreen shrubs or small trees to 25′ with small leaves. Grows in mild climates.

Plant in spring or early fall in partial shade, though box does well enough in full sun. Grow in average, well-drained soil with some humus added. Set the plants slightly deeper than they formerly grew and water copiously for several weeks. If plants are large, shade them a little from the sun for several weeks. Water established plants in dry weather, and water heavily in the fall before the ground freezes. Fertilize every year or two. Trim in August. In colder areas, build a frame over the plants and cover with burlap to keep off the snow, which may break the branches. In cold, windy areas, surround exposed plants with burlap or evergreen boughs to break the wind. If leaves develop holes in summer, spray with malathion for leaf miners.

BOYSENBERRY Bramble fruit much like loganberries with very large, dark-red fruits. Grows in warm climates; is killed by cold.

Grow like blackberry (which see). However, plants should be spaced 8′ apart in the row and should be trained to horizontal wires strung between posts.

BRACKEN *Pteridium aquilinum*, brake. Coarse fern to 3′ with triangular fronds and creeping rootstocks. Grows almost everywhere.

For how to grow, see fern. This is a tough, undemanding plant that will grow in very poor soil and in full sun or shade.

BRAIN CACTUS *Stenocactus.* Round desert cacti to 6″ with wavy ribs which make the plants look wrinkled, much like a brain. Grows indoors; also in hot climates outdoors.

For how to grow, see cactus, desert.

BRAZILIAN NIGHTSHADE *Solanum seaforthianum,* potato vine. Woody vine about 25′ tall with clusters of violet flowers and then inedible yellow berries. Grows in warmest climates.

This plant propagates itself by seeds. Plant rooted specimens in sun in well-drained, average soil in spring or fall. Water in dry weather. Feed 3–4 times a year. Prune annually after flowering (which continues on and off through much of the year). Provide a trellis for vine to twine up on.

BREATH OF HEAVEN *Diosma ericoides.* Shrub 4′ tall with aromatic foliage and tiny white flowers. Grows indoors; also outdoors in warmest climates.

Indoors. Pot in two parts loam, two parts humus and one part sand. Water well when soil feels dry. Grow in a south window. Fertilize every two months during fall and winter. Prune hard in early spring after flowers fade.

Outdoors. Plant in spring or fall in sun in light, humusy soil. Water in dry weather. Fertilize in late winter. Prune hard after flowering.

BRISBANE BOX *Tristania conferta.* Handsome evergreen tree to 100′ with glossy leaves and reddish bark. Grows in subtropics.

Plant in spring or fall in sun in average soil. Water deeply about once a month in dry weather. Fertilize every 3–4 years.

BRIZA Quaking grass. Annual grass to 2′ with slender stalks and flattened, pendulous bronze heads. Decorative when dried. Grows almost everywhere.

Sow seeds in spring where plants are to grow. Give sun and an average soil. Water in very dry weather. To dry, hang stalks head down in a shady place.

BROCCOLI *Brassica oleracea italica.* A member of the cabbage family with green, cauliflower-like heads on 2′ stalks or branches. Grows best in cool climates.

Broccoli is happy in average soil which is improved with nitrogen-rich commercial fertilizer at planting time and monthly thereafter. You can grow two crops a year. for the first crop, sow seeds indoors about 10 weeks before setting-out time, and transplant to the garden when danger of hard frost is past. Or buy plants and set them out after last hard frost. For the second crop, sow seeds directly in the garden in late June or early July. Plant in rows 2′ apart with 2′ between plants. Broccoli needs full sun. Keep cultivated. Water during dry spells. Cut heads with about 4″ stems while flowers are in tight bud. New heads will develop on side branches.

BRODIAEA Triplet lily, Ithuriel's spear, snake lily, spring starflower, blue dicks, pretty face, wild hyacinth. *Brodiaea* is sometimes identified as *triteleia*. Plants to 2′ growing from corms and bearing funnel-shaped flowers in various colors in spring. Snake lily climbs by twining to a height of 8′. Grows in California and the south.

Plant corms in fall in a sunny spot. The corms should be set 3″ deep and about 6″ apart in well-drained soil that is both sandy and humusy. Give a light dose of plant food every other spring. Water moderately while making growth, but gradually reduce the supply after flowering.

BROOM *Cytisus*. Deciduous or evergreen shrubs to 10′ with pretty spring flowers in yellow, white, purple or pink. Grows in mild and warm climates. See also *genista* and Spanish broom.

Buy potted plants. Do not try to move established specimens. Plant in spring or fall in sun in average, well-drained soil. Water in dry weather. Fertilize every year or two in the spring. Prune after blooming.

BROWALLIA Amethyst, sapphire flower. Annuals to 2′ with blue flowers. Grows almost everywhere; also indoors. *B. americana* is the more common garden plant; *B. speciosa major*, the more common house plant.

Outdoors. Sow seeds indoors eight weeks before last frost, or sow outdoors where plants are to grow when weather warms. Grow plants in full sun in average soil. Water in dry weather. Fertilize when young plants are established.

Indoors. Sow seeds outdoors in late summer and transplant twice into pots of increasing size (up to 4″ maximum). Bring in before frost and place in a sunny, very cool (about 60°) window.

BRUCKENTHALIA SPICULIFOLIA Spike heath. Ten-inch, tufted, evergreen shrub with needle-shaped leaves and pink flowers. Grows in mild and warm climates.

Propagate by cuttings. Plant in spring or fall in a sunny spot in sandy soil. In cold climates, mulch with leaves or straw in winter. Fertilize every year or two. Water in dry weather.

BRUNFELSIA One species known as yesterday-today-and-tomorrow plant; another as lady of the night. Evergreen shrubs to 5′ with fragrant, tubular flowers that are white or violet. Grows outdoors in warmest climates; also a house plant.

Outdoors. Propagate by young stem cuttings in summer. Plant in fall in partial shade or sun in good, acid soil containing extra peatmoss. Water in very dry weather. Fertilize in winter before growth begins. Prune as necessary to control growth in late spring or summer.

Indoors. Pot in general-purpose potting soil. Grow in an east or west window. Keep moist and at a night temperature no lower than 55°. Fertilize and prune as above. Do not repot often.

BRUSSELS SPROUTS *Brassica oleracea gemmifera*. Erect 2′ member of the cabbage family producing in the fall a dense cluster of tiny, edible buds. Grows in cool climates.

Sow seeds outdoors in flats or a seedbed in not-too-rich soil about June 1. Transplant into full sun in average soil in the garden not later than August 1. Space plants 18″ apart in rows 2′ apart. Sidedress with nitrogen-rich fertilizer as soon as plants are established. Keep soil cultivated and watered during dry spells throughout the growing period. When sprouts begin to form, break off the lowest leaves of the plant. Cut developed sprouts off the stems with a knife, working from the lowest up.

BRYOPHYLLUM Succulents to 5′, which produce new plants on their leaves and have colorful flowers. One popular species known as air plant. Best grown indoors.

For how to grow, see cactus, desert. Though bryophyllum must be protected from cold, it likes to be moved outdoors in summer. The little plants can be removed from the leaves and planted in regular cactus soil to make large plants.

BUCKTHORN *Rhamnus*. Different species called coffeeberry; Indian cherry or yellow bush; cascara sagrada, bearberry or chittamwood; Hart's thorn, waythorn or rhineberry. Deciduous and evergreen shrubs and trees to 40′ with bright-green leaves and small fruits. Different species grow in different parts of the country.

Plant in spring or fall in average soil in sun or partial shade. Water in dry weather.

BUDDLEIA Butterfly bush, summer lilac. Deciduous shrubs to 10′ with very showy spikes of violet, white or pink flowers in spring or, usually, in summer. Grows best in mild and warm climates.

Propagate by tip cuttings of side shoots in summer. Plant in spring or fall in sun in well-drained, average soil to which some humus has been added. Water in dry spells. Fertilize in spring. If plants are killed by cold, they usually come back quickly. Most plants can be pruned rather severely in early spring if they become too large (however, *B. alternifolia* is pruned right after flowering).

BUGBANE *Cimicifuga*, snakeroot, black cohosh, rattlesnakeweed. Perennials to 6′ with handsome foliage and white flower spikes in summer and fall. Grows in warm climates.

Divide plants in early spring, and plant in partial shade. Provide a rich soil with plenty of humus. Water regularly. Fertilize in spring. Topdress with humus in the fall.

BULBOCODIUM Red crocus, meadow saffron. Crocus-like, spring bulb to 4″ with yellowish flowers. Grows best in temperate climates.

Plant 3″ deep and 3″ apart in average, well-drained soil in a sunny spot in the fall.

BUNCHBERRY *Cornus canadensis*. Six-inch plant with a whorl of dogwood-like leaves, small white flowers and red berries. Grows in cold, dampish climates.

Transplant from the wild in early spring or fall. Be careful to take a large chunk of soil with the roots. Plant in shade in acid soil that contains considerable humus and holds moisture.

BURNET *Sanguisorba minor*. Perennial to 1′ with fern-like leaves which are used in salads and for other flavoring. Grows almost everywhere.

Divide plants in spring and plant

in sun in average soil. Water in dry weather. Fertilize in spring.

BUTTERFLY WEED *Asclepias tuberosa*, orange milkweed. Three-foot perennial with showy, flat-topped clusters of bright orange flowers. Grows mainly east of the Rockies.

Sow seeds in the fall where they are to grow or in a seedbed. Thin or transplant when seedlings are developed. Give full sun and average, light, well-drained soil. Do not feed. Little watering is required except in severe drought. Don't disturb established plants. Cut seed pods before they mature.

BUTTON BUSH *Cephelanthus occidentalis*, button willow. Deciduous shrub to 20′ with glossy foliage and fragrant, white, ball-like flowers in late summer and fall. Grows in mild and warm climates.

Plant in spring or fall in sun or partial shade in average soil which holds moisture. Grows very well in swampy areas. Fertilize every 2–3 years.

BUTTONWOOD *Platanus occidentalis*, sycamore. Huge deciduous tree to 140′, with handsome, toothed leaves and mottled, light and dark bark. Grows widely east of the Mississippi and into Texas.

Plant in spring in full sun in an open situation. The soil should be moist and of above-average quality (buttonwoods grow best in river bottoms). Foliage may be attacked in the spring by a fungous disease which makes young leaves turn brown, curl up and drop off. To prevent this, spray with Bordeaux mixture when leaves are beginning to open. Except for this, the tree pretty well takes care of itself, though in early life it will appreciate some fertilizer every 2–3 years.

CABBAGE *Brassica oleracea capitata*. Leafy vegetable grown in cool climates. Different varieties are required for spring and fall crops.

Grow in full sun in average soil enriched at planting time and one month later with commercial fertilizer rich in nitrogen. For spring crop, sow seeds indoors about eight weeks before plants are to be set out. Don't fertilize the seedbed. Keep cool. Transplant to garden after last frost. Space plants 12″–15″ apart in rows 18″ apart.

For fall crop, sow seeds outdoors either in flats or in a seedbed about June 15, and transplant into garden by August 1. Space plants 2′ apart in both directions.

Both types of cabbage require regular cultivation and frequent watering during dry weather. To prevent damage from cutworms and root maggots, wrap stems of newly set out plants with 2″ strips of paper or lay disks of tarpaper on the ground around the stems. Dust plants regularly with rotenone to control the cabbage worms that eat the leaves, and apply nicotine to control aphids. Do not plant in the same bed year after year.

CACTUS, DESERT A succulent type of plant which stores water and lives in dry, usually hot areas. There are many different kinds, and contrary to popular opinion, not all are prickly. Generally speaking, all desert cacti are grown in the same way, though exceptions may be noted under particular plants listed in this book.

Outdoors. Plant in spring in a sunny location that is raised slightly to prevent unnecessary frost damage to plants. The ideal bed consists of 3″ or more of coarse stones, then 3″ or more of smaller gravel, then 12″ or more of soil composed of one part loam, one part humus and one part coarse sand.

Do not water plants before planting, and make sure the soil is dry. Then place plants in position and firm soil around roots. If plants cannot support themselves, stake them. Do not water for 3–4 days; then apply just enough to dampen the soil. Keep soil more dry than damp (but not completely dry) for the next two months. This care is necessary because newly planted cacti rot easily.

Once a cactus plant is established, regular watering is required when the plant is growing in spring and summer. Apply the water when the soil dries out on top, and soak the soil thoroughly and deeply. In fall and winter, when the cactus is resting, water should be applied only to prevent shriveling; unless you live in a totally arid climate, normal rainfall should supply all the moisture required.

In spring, when the plant starts to grow, give it a small amount of general-purpose fertilizer or bone meal mixed in a little humus. Keep area around plant weeded. Cut off damaged or dead growth with a sharp knife, and dust large wounds with powdered sulfur. Protect plants against a hard freeze by covering with newspapers, burlap or a paper bag.

Indoors. Plant in a clean pot that does not dwarf the plant. The pot must have 1″ or more of coarse drainage material. Use the soil mixture described for outdoor planting.

Handle as above, but observe these additional points: Grow in a sunny south window and turn the pot occasionally so that all sides of the plant are exposed to the sun. Ventilate the room in which the plant is growing frequently. Move plant outdoors in summer, but do not expose to the full sun for a week or so. During winter, keep the plant in the coolest (down to about 40°) room available. Repot plants in fresh soil about every two years, or when they obviously become root-bound.

Propagating. The easiest way to propagate cacti, no matter where they are grown, is to take cuttings in spring or summer as soon as the plants start into active growth. Use plump leaves or stems. Cut off with a sharp knife and dust the wound on the plant (but not on the cutting) with powdered sulfur. Keep the cutting in a cool, dim, ventilated place for about ten days. Then place the cut end of the cutting in almost dry sand that is 2″–3″ deep. Insert in the sand only enough for it to stand upright. (To avoid planting long cuttings too deeply, tie them to stakes.) Keep in a warm, lightly shaded, airy place and water sparingly until roots develop. Then transplant into regular cactus soil.

Insects and diseases. To control scale on cacti, spray with an oil emulsion spray several times. To get rid of mealybugs, spray with a strong spray of water or pick off individual bugs with a cotton swab dipped in alcohol. If strange swellings appear on the roots of plants, trim the roots back, let dry for several days and then repot in fresh soil. Sterilize the old soil before using it for other plants. Spray with nicotine sulfate solution to control aphids and red spider. Kill snails and slugs attacking plants with a poison bait. Cut out decay spots with a sharp, sterile knife, dust with sulfur and allow to form a callus; then replant.

CAJEPUT *Melaleuca leucadendron,* paper bush, punk tree. Evergreen shrub or small tree to 20′ with shredding bark, narrow, tapering leaves, and white flower spikes. Grows in warmest climate.

Plant in early spring or fall in a sunny spot. The tree is extremely adaptable, grows in almost any soil, withstands wind, salt spray and drought. But for best growth, water it regularly and fertilize every 3–5 years.

CALADIUM Tuberous plants to 18″, grown for their handsomely colored leaves. Popular outdoors in warm climates. Also a good house plant.

Outdoors. Plant in beds or pots when soil is warm and danger of frost is past. Tubers should be 1″–2″ deep and 6″–10″ apart in rich, humusy, well-drained, acid soil in filtered sun. Fertilize every 2–3 weeks. Keep soil moist but not soggy. Spray foliage with water in dry weather. When foliage dies down in fall, withhold water and bring tubers indoors to dry off. They can be stored in their pots or in boxes of sand or vermiculite. Storage temperature should be about 60°.

Indoors. Plant 1″ deep in two parts loam, two parts humus and 1 part sand. Keep moist. Feed every 2–3 weeks. Grow in a warm east or west window. Spray with water 2–3 times a week. Gradually withhold water when growth stops, and store tubers in their pots in a warmish, dark place. Start into growth again about March.

To propagate caladiums, cut the tubers into sections, each containing one or more eyes. Dust cuts with sulfur. Plant in a moist sand-peat-moss mixture. Move into pots when rooted.

CALAMONDIN Evergreen tree to 25′ with white flowers and fruits like a small tangerine but hardly edible because of their acidity. Grows mainly in Florida.

Grow like orange (which see).

CALANTHE Mostly deciduous, terrestrial orchids to 3′ with clusters of flowers usually white or rosy. A house plant.

For how to grow, see orchid. Plant calanthe in well-drained pots in general-purpose potting soil (two parts loam, one part humus, one part sand). Repot every two years just before growth starts in spring. While making growth, water when soil feels dry; but when plants begin to lose foliage in late summer, reduce supply considerably; and when flowers are gone and plants go into their rest period, move the plants in their pots into a dim place and water lightly to keep pseudobulbs from shriveling.

Don't spray tops of calanthe during flowering, but keep pots standing on wet pebbles at all times. Don't move plants outdoors in summer except in areas where nights are very warm.

CALATHEA Three-foot perennials with colored leaves. Grows indoors; outdoors only in extreme southern Florida.

Pot in two parts loam, two parts humus and one part sand. Keep moist but not wet—water when soil feels almost dry. Spray with water once or twice a week. Fertilize every month with liquid plant food. Grow in a warm east or west window. Spray with malathion to control spider mites. Repot in the spring.

CALCEOLARIA Slipperwort. Plants to 3′ with slipper-shaped, yellow and brownish flowers. Mainly a florist's plant.

CALENDULA Pot marigold. Annual to 2′ with yellow or orange flowers. Grows almost everywhere.

For earliest bloom, sow seeds indoors 4–6 weeks before last frost; or sow outdoors after last frost where plants are to grow. In hot climates, make a second sowing outdoors about June 21 for flowers well into the fall. Calendula needs full sun, average soil, normal moisture, one dose of fertilizer in the spring when the small plants are well established.

CALICO FLOWER *Aristolochia elegans.* Slender evergreen vine to 20′ with cornucopia-like, white-with-purple flowers. Subtropical.

Propagate by self-sown seedlings, or plant seeds yourself in early spring and provide full sun until summer, when light shade is preferred. Grow in a moist, humusy, fertile soil. Keep watered except in winter, when little is needed. Fertilize in late winter. Provide support vine can twine up on, and help it to get started. Cut back hard after flowering and replenish top soil.

CALIFORNIA LAUREL *Umbellularia californica,* California bay, Oregon myrtle. Evergreen tree to 85′, but about a third as tall in cultivation. Has aromatic leaves, yellow flower clusters and purplish fruits. Grows in California and Oregon.

Plant in spring or fall in sun or partial shade in average soil. Water deeply in dry weather, because the laurel requires a good amount of moisture. Fertilize every 2–3 years. Prune in winter as necessary. Tree will grow more shrub-like in the garden.

CALIFORNIA POPPY *Eschscholtzia californica.* Spreading annuals to 20″ with abundant cup-shaped flowers in many colors. Grows almost everywhere.

Sow seeds outdoors where plants are to grow as soon as soil can be worked. Seeds may also be sown outdoors in very late fall. Do not transplant, but thin to about 9″ apart and fertilize at that time. These poppies need sun and average soil. Water in very dry weather.

CALLA LILY *Zantadeschia.* Tuberous plant to 30″ with funnel-shaped, upright flowers in white, yellow or pink. Grows outdoors in warm climates; can be grown outdoors in cold climates if treated as a tender bulb. Also a house plant.

Outdoors in warm climates. Plant the tubers in filtered sun in fall. Set 3″ deep (6″ in areas where ground may freeze) and 6″–12″ apart. The soil must be rich, very humusy and acid. Keep moist but not soggy at all times. Fertilize every month while plants are making growth. Divide plants after flowering when they become crowded.

Outdoors in the north. Plant outdoors in spring when soil is warm and frost danger is past. Grow as above. In the fall, dig and store tubers like tuberous begonias (which see).

Indoors. Plant tubers ½″ deep in 6″ pots containing general-purpose potting soil enriched with extra humus. Keep in a cool, dim place and water sparingly until growth starts. Then move into a brighter, warmer place; keep moist, and fertilize every three weeks. When flower stalks develop, move into a south window at not more than 65°. Continue to water and feed. In spring, when weather is warm, unpot the plants and set in the garden in partial shade. Continue watering and feeding. Then, in late summer, gradually let foliage die down; dig up tubers, and store in a cool, dry place for two months.

CALLIANDRA Powder puff, flame bush. Evergreen shrub to 12′ with clusters of flowers in red, pink, or white. Grows in warmest climates.

Propagate by air layering. Plant in spring or fall in a sunny spot where plant will have plenty of room to spread. Soil can be of average quality, but acid. Water in very dry weather, especially when plant is young. Fertilize about October and again about January. Thin out branches in summer.

CALLICARPA Beautyberry. One species also called French mulberry. Deciduous shrubs to 10′ with violet or white, berry-like fruits in fall and white, blue or pink flowers in summer. Grows in warm climates.

Propagate by stem cuttings in spring. Plant in spring or fall in partial shade in soil with ample humus and loam. Fertilize every spring. Water in dry weather. Prune hard in later fall. If plants are killed by frost, they will probably put out new growth from the roots.

CALLIOPSIS Coreopsis (which also see). Annual version of coreopsis, to 36″, with daisy-like, yellow, orange or reddish flowers. Grows anywhere.

Sow seeds indoors 6–8 weeks before last frost, or sow outdoors after frost. Grow in sun in average, well-drained soil. Fertilize when young plants are established. Water in dry weather.

CAMASSIA Bulbous plant to 2′ with large spikes of flowers, mainly blue, in spring. Grows best in California but grows also in other temperate regions.

Plant bulbs in the fall in sun or light shade. The soil should be well-drained, rich, somewhat sandy, not too dry. Set bulbs 4″ deep and about 5″ apart. Do not disturb thereafter.

CAMELLIA Evergreen shrubs or trees to about 30′ with handsome, waxy flowers in reds and other colors in early spring or late winter. Grows in warm climates. People also try to grow camellias indoors but without success unless they have a sunny room where the temperature stays no higher than 50°.

Propagate by cuttings of mature stems. Plant in spring in light shade. The planting hole should be deeply dug (to 3′) and filled with 4″–6″ of coarse drainage material, then with a mixture of one part loam, one part peatmoss and one part sand to which aluminum sulfate is added to increase the acidity. Set plant at the level it formerly grew and keep well watered for several weeks. Thereafter, water regularly so that soil is almost constantly moist. Spray foliage with water daily in summer. Apply a cupful of 4–8–6 plant food annually in early spring; make a second application in June or July if the soil is poor. To maintain acidity and hold in moisture, mulch soil with oak leaves, pine needles or (less good) peatmoss. In alkaline soil, the addition of some aluminum sulfate is probably needed. Spray plants with oil emulsion spray before growth starts (and if necessary, afterwards up to the onset of hot weather) to control mites and scale. If leaves on a branch turn brown and curl under in spring, cut out the branch completely and burn it to control dieback.

CAMPANULA Bellflower, Canterbury bells, peach bells, bluebells of Scotland, Coventry bells. Large genus of annuals, biennials and perennials from 6″ to 5′ tall, with delightful, bell-shaped flowers in blue, white, rose and lavender. Grows almost everywhere. *C. isophylla* (star of Bethlehem) is also a house plant.

Outdoors. Campanulas grow well in average, well-drained soil that is enriched with general-purpose fertilizer in early spring and about twice thereafter during warm weather. Grow in full sun or light shade. Water regularly. Stake tall varieties. Keep dead flowers picked off. Protect plants in very cold climates in winter. Divide perennials every 2–3 years.

Start annuals from seed sown indoors about 8–10 weeks before the last frost of spring. Start biennials from seed sown in summer, and transplant to their permanent locations in the fall. Handle perennials in the same way, or divide plants in early spring.

C. isophylla indoors. Propagate by division in the spring. Pot in general-purpose potting soil. Grow in a cool south window. Keep soil moist and apply fertilizer every 3–4 weeks. Move outdoors in summer. When plant is through blooming in late fall or early winter, move to a very cool (about 50°), bright room and water just enough to keep from shriveling. Thin out old stems. Then in spring divide or repot in new soil, and start all over again.

CAMPHOR TREE *Cinnamomum camphora.* Wide-spreading evergreen tree to 40′ with camphor-scented foliage. Grows in warm climates.

Plant pot-grown specimens in spring in sandy loam. The tree needs sun and an isolated position, because its roots spread far and wide and make cultivation of anything else difficult. It also casts very heavy shade. Water in dry spells. Fertilize every 3–4 years.

CANDYTUFT *Iberis.* Spreading plants to about 1′, covered with flowers of various colors. Most species are annuals. *I. sempervirens* is an excellent evergreen perennial with white spring flowers. Candytuft grows almost everywhere.

Annual. Sow seeds where they are to grow as soon as soil can be worked. Plants need full sun, average soil. Water in dry weather. Fertilize lightly when plants are thinned. Make several sowing of seeds at 2–3 week intervals for a long season of bloom.

Perennial. Sow seeds in a seedbed any time until late June; move into the garden in the fall. Or divide established plants in the fall and set in their permanent locations. Give sun, water in dry weather, average soil. Fertilize in early spring and again after flowering.

CANNA Tuberous plants to 6′ with clusters of large, showy, summer flowers in various colors. Grows almost everywhere.

Propagate by division of the roots in spring. Each section should have 1–2 eyes. For an early start in cold climates, plant the roots in pots or flats indoors and transplant outdoors when danger of frost is past. Or you can plant the roots directly in the garden when the weather is reliably warm.

Cannas need full sun and deeply dug, well-drained, rich soil. Plant roots 4″ deep and 18″ apart. Cultivate and water regularly. Fertilize every 3–4 weeks with general-purpose plant food. In the fall, when the plants are killed back by frost, dig up the roots and cut off all but about 6″ of the growth. Dry off and store like dahlias (which see) in peatmoss in a dry, cool place.

CANTALOUPE *Cucumis melo,* muskmelon. Wide-spreading, orange-fleshed melon with ribbed, netted rind. Needs high temperature but can be grown almost everywhere.

Cantaloupes need full sun, good air circulation. The soil should be well drained, sandy and humusy. Dig it deeply and mix in lime and general-purpose fertilizer rich in potash. As soon as danger of frost is past and soil is warm, sow a half dozen seeds 1″ deep in 1′ diameter circles (hills) which are placed 4′–6′ apart. Thin to one strong plant per hill. Dust plants every week until they have flowered with rotenone. Cultivate lightly but thoroughly to keep out weeds. Water frequently in dry weather. Fruits are ready to pick when stem comes loose with a slight touch of the finger. In the fall, burn all plant debris.

In regions where the summers are short, you can get a head start with melons by sowing seeds indoors and transplanting outdoors after all danger of frost is past. Or you can sow seeds outdoors and cover plants with Hotkaps.

Note that cantaloupes do not change flavor if planted next to cucumbers.

CAPE COWSLIP *Lachenalia.* Bulbous plants to 1′ with spikes of tubular flowers that are yellow and red. Mainly a house plant.

Grow like freesia (which see). After flowers fade, keep foliage growing as long as possible by watering regularly. Then when foliage begins to die down, reduce water. When foliage disappears, stop watering altogether and store the bulbs in their pots in a dry, cool place.

CAPE HONEYSUCKLE *Tecomaria capensis.* Evergreen vine to about 12′ with showy clusters of orange-red flowers during much of the year. Grows in subtropics.

Propagate by layering. Plant in rich, moist, sandy soil. Water regularly. Fertilize in spring and again in summer. Grow in the sun. Provide a strong support and train the branches on it. Prune after flowering. By hard pruning, this plant can be grown as a shrub only a few feet tall.

CARAWAY *Carum carvi.* Biennial to 2′ with carrot-like leaves and aromatic seeds. Grows almost everywhere.

Sow seeds in spring in a sunny spot where plants are to grow. Soil should be of average quality. Water in dry weather, and fertilize when seedlings are thinned to 6″ apart. Fertilize again the following spring. After flowering in the second year, seeds will develop and can be harvested.

CARDINAL'S GUARD *Pachystachys coccinea.* Sometimes identified as *jacobinia.* Perennial to 5′ with spikes of tubular, red flowers in summer and fall. Grows in warmest climates.

Propagate by cuttings in early spring. Plant in early spring in average soil in partial shade. Water in dry spells. Fertilize in spring and again about two months later.

CARDOON *Cynara cardunculus.* Perennial vegetable to 6′, grown for its blanched leafstalks, which are boiled or eaten in salad. Grows in warm climates.

Cardoon grows like artichokes, to which it is related. See artichoke. Be sure to water well in dry weather. About September 15, tie the leaves together and wrap with heavy paper for a month to blanch the stalks.

CAROB *Ceratonia siliqua,* St. John's bread. Evergreen tree to 40′ producing large edible pods. Grows in warmest climate.

Plant only young specimens that come with a good ball of earth. Both male and female trees are needed for fruit. Plant in early spring or fall in sun in average soil. If grown in arid regions, water deeply several times during the summer to improve yields of pods. Also to improve yield, fertilize every 2–4 years.

CAROLINA JASMINE *Gelsemium sempervirens.* Woody vine to 20′ with shiny evergreen foliage and a profusion of yellow flowers in early spring. Grows best in the southeast.

Propagate by stem cuttings. Plant in average, well-drained soil with extra humus and some fertilizer. Grow in sun or light shade. Keep well watered. Spray foliage with water occasionally. Mulch the soil to hold in moisture and keep roots cool. Fertilize early every spring. Grow on a sturdy support and steer the stems where you want them to go while plant is young. Remove dead and damaged wood and prune to shape right after flowering.

CARPENTERIA CALIFORNICA Evergreen shrub to 7′ with green and white leaves and fragrant, white, anemone-like flowers in summer. Grows in California.

Propagate by layering. Plant in spring or fall in sun in average soil that is sandy and well drained. Water deeply, but only about once a month, in dry weather. Fertilize in spring. Prune in winter if necessary to control shape. Spray with malathion should aphids get started.

CARPET GRASS *Lippia canescens.* Creeping plant only a few inches high. Used as a groundcover in warm climates.

Divide into small sections and plant 1′ apart in spring in sun in average soil. Water well to establish plants; then in dry weather (carpet grass requires less moisture than ordinary grass). Fertilize lightly every spring. Mow as you please.

CARROT *Daucus carota.* Familiar root vegetable to 20″. Grows almost everywhere.

Carrots require full sun and light, rich soil that has been dug deeply, thoroughly pulverized, cleared of stones and improved with lime and fertilizer. Sow seeds in early spring in rows 1′ apart and thin plants when they are about 2″ tall to 2″

apart. Make additional sowings when the one preceding is well up. Keep soil lightly cultivated. Water regularly. The faster carrots grow, the more tender they are. Dig them before they are full grown.

CARYOTA One species called fishtail palm. Palm trees to 70′ with fishtail-shaped leaflets. Subtropical.

For how to grow, see palm. This genus of palm needs plenty of water, especially during spring and summer. The fishtail palm dislikes hot, dry winds.

CASABA MELON *Cucumis melo.* Large melon with tough yellow rind and white flesh. Requires about 120 days to mature; is therefore best adapted to warm climates.

Grow like cantaloupe (which see).

CASSIA Senna. Trees and shrubs to 50′, some evergreen, some deciduous, usually with yellow flowers, but one called pink shower has pink flowers. Grows in warmest climates. There is also a small perennial cassia sometimes grown in the east.

Plant in spring or fall in sun in average, well-drained soil. Water in dry weather. Fertilize every year or so. Prune after flowering. Spray with rotenone to control caterpillars; any other insecticide will probably kill the leaves.

CASTOR OIL PLANT *Ricinus communis.* Annual to 15′ with big beautiful leaves and poisonous seeds. Grows almost everywhere.

Sow seeds indoors about six weeks before last frost, or sow outdoors where plants are to grow when weather warms. Give full sun and a rich, well-drained soil. Fertilize frequently and water regularly to encourage fast growth.

CASUARINA Australian pine, beefwood, she oak. Curious leafless trees to 60′ with jointed twigs. Grows in warmest climates.

Plant in spring or fall in sun in average soil (it is often grown in the sandy soils found along sea coasts). Requires little attention. If used in a windbreak, prune in winter.

CATALPA Cigartree, Indian bean, bean tree. Deciduous trees to 60′ with handsome flower clusters in late spring and long, narrow seed pods. Grows east of the Mississippi.

Plant in early spring or fall in full sun. Thrives in average soil and makes very fast growth without help. But you'd better water in very dry weather.

CATANANCHE Blue succory, cupid's dart. Slim perennial to 2′ with crisp, blue, ray flowers in summer. Grows almost everywhere.

Divide and plant in spring in a sunny position in average, well-drained, preferably sandy soil. Water only in very dry weather. Apply fertilizer in spring.

CATSCLAW CREEPER *Doxantha unquis-cati,* yellow trumpet vine. Rampant evergreen vine to 40′ which sends out unbranched stems from the base and bears large, yellow, trumpet-shaped flowers in spring. Endures considerable cold but only for brief periods, is therefore best grown in warm climates.

Propagate by seeds or tip cuttings in spring. Because this creeper can take over your place, plant it in full sun in a spot where it will not become a nuisance. It grows in indifferent soil; tolerates drought, some wind and dry, desert air. Provide strong support which the tendrils can grasp. Prune back very hard—even to the base—after flowering, before unsightly seed pods develop.

CAT TAIL *Typha latifolia,* flag mace, reed mace. Reedy perennials to 9′ with dense brown spikes of flowers. Grows almost everywhere.

Divide roots in spring and plant in sun in average soil which is wet or under a few inches of water.

CATTLEYA Most familiar of the orchids. Mostly epiphytic, evergreen plants to about 2′ with huge, usually purple or white flowers at different times of the year, depending on the species. A house plant.

For how to grow, see orchid. Repot *cattleyas* soon after they have flowered, or pack more osmunda in around them.

CAULIFLOWER *Brassica oleracea botrytis.* A member of the cabbage family with large white or purple head. Does not tolerate heat or drought. Best grown in most parts of the country as a fall crop.

Cauliflower needs moderately rich soil which is further improved with fertilizer at planting time. Sow seeds in flats or a seed bed about June 1 and transplant to the garden by August 1. Plants should be 2′ apart in all directions. Keep thoroughly cultivated and water heavily in dry weather. When heads begin to form, sidedress with general-purpose fertilizer. When heads are about the size of an egg, fold several of the large outer leaves up over them and tie the leaves loosely together. Cut the filled out heads before frost.

CEANOTHUS Wild lilac. Sizable family of evergreen and deciduous shrubs from 4′ to 35′, with clusters of white and blue flowers. Grows on west coast. *C. americanus,* also called New Jersey tea, Indian tea or redroot, is a 3′ shrub occasionally grown in the east.

Plant in spring or fall in sun in light, well-drained soil. Water thoroughly but not often in dry weather. Fertilize in winter. Whether and how much to prune depends on the species, since some require it to prevent legginess while others react badly; so ask the dealer from whom you buy plants. Many species are short-lived.

CEDAR *Cedrus.* The true cedars include the Atlas cedar, deodar and cedar of Lebanon. Conifers to 100′ of somewhat varying habit, with erect cones and needles in clusters. Grows in warm or mild climates. See also arborvitae and juniper.

Plant in average, well-drained soil in spring or fall. Grow in the sun and allow plenty of space for maximum development. Water well until tree is established. Fertilize every year or two while tree is young.

CELERIAC *Apium graveolens rapaceum,* knob celery. Plant to about 2′, raised for its large, white, celery-flavored root. Grows in cool climates.

Grow like celery (which see). Space 6″–8″ apart in rows 2′ apart. Do not blanch. Remove lateral roots that develop, close to the crown of the root.

CELERY *Apium graveolens.* Vegetable to 28″ grown in cool climates, but may also be grown in winter in the south as a spring crop. Yellow type is self-blanching. Green type is used green or may be blanched.

Celery grows best in drained muck lands, but can be grown in any fertile soil which is deeply dug and mixed with lime, general-purpose fertilizer rich in phosphorus and potash, and humus in great quantities. For a fall crop, soak seeds in water for 24 hours, then sow in a flat about the middle of May. Transplant seedlings into another flat or seedbed when they are about 2″ tall. Keep soil moist. Transplant into the garden about the middle of July when plants are 3″–4″ tall and stocky.

For a spring crop, sow seeds indoors about eight weeks before plants are to be put in the garden. Transplant into flats or the coldframe when 2″ tall.

Whenever celery is started, the young plants should be set 8″ apart in rows 2′–3′ apart in a sunny location. Water the plants before they are moved; also the garden soil. As the plants grow, keep them carefully weeded and never let the soil dry out. When applying water, soak the ground; get as little as possible on the plants.

It is not necessary to blanch celery. But if you wish to do so, the safest method to follow is to stand 1′ wide boards on edge on both sides of the plants in a row when the plants are mature. The boards should not cover the leaves. Slant them toward each other and hook them together at the top so that very little light reaches the celery stalks.

CELTUCE *Lactuca sativa asparagina.* Member of the lettuce family, to 18″, raised primarily for the fleshy central stem which is eaten raw or cooked like celery. Grows best in cool climates.

Celtuce is raised on average soil which has been limed. Sow seeds in rows 15″ apart in a sunny location as soon as the soil can be worked in the spring. Thin plants to 10″. The leaves at this stage can be used like lettuce. Harvest plants when stems are $\frac{3}{4}$″–1″ in diameter. While plants are growing, water in dry spells. Sidedress with fertilizer at time of thinning.

CEPHALOTAXUS Plum yew. Graceful evergreen trees to 30′ with needled foliage and plum-like green or purple fruits. Grows best in mild and warm climates near the seacoast.

Plant in spring or early fall in lightly shaded and sheltered locations. Grow in well-drained, average soil which contains a fair amount of sand. Water well until plants are established. Fertilize every year or two while trees are young.

CERASTIUM Snow in summer, starry grasswort. Spreading perennials less than a foot tall with white flowers in spring. Grows in temperate climates.

Sow seeds or divide plants in spring. Grow in sun in average, dry soil. Fertilize lightly in early spring. Water in very dry weather.

CERATOSTIGMA Plumbago (see also entry under plumbago). Shrubs to 4′ with clusters of blue, phlox-like flowers in summer and fall. Grows in mild and warm climates, especially California.

Plant in spring or fall in sun or light shade in average soil. Water in dry weather, especially if growing in the sun. Fertilize in spring. Cut back hard after flowering.

CEREUS Ribbed desert cacti to 40′ resembling candelabra. Grows outdoors where temperature does not fall below 20°. Also grown indoors.

For how to grow, see cactus, desert.

CERCOCARPUS Mountain mahogany. Graceful, usually evergreen shrubs or trees to 20′ with inconspicuous flowers and small fruits with feathery tails. Grows mainly in California.

Plant in spring or fall in sun in average, well-drained soil. Water deeply but not too often in dry weather.

CEROPEGIA One species (*C. woodi*) known as rosary vine. Mostly trailing succulents, with stems several feet long and with funnel-shaped, often multi-colored flowers. Grows indoors.

For how to grow, see cactus, desert. Grow in light shade and water regularly only in summer. In winter, keep quite dry. Propagate tuberous types like the rosary vine by division of the tubers; others by cuttings.

CESTRUM One species (*C. nocturnum*) called night-blooming jasmine; another (*C. diurnum*) called day jasmine. Evergreen shrubs to 12′ with very fragrant white flowers. Subtropical.

Propagate by stem cuttings in late winter. Plant in fall in average, humusy soil in partial shade. Water regularly. Fertilize in fall and a couple of times thereafter. Cut back hard after flowering stops (or if you have a plant that blooms much of the year, cut back when it slows down a little). Pinch stem ends occasionally to promote bushier growth.

CHAIN FERN *Woodwardia.* Fern with narrow fronds to 3′. Grows best in eastern coastal regions.

For how to grow, see fern. The chain fern needs a moist, humusy soil. Grow in very light shade or partial sun.

CHAMAEDOREA Large genus of palms, some quite small, others up to 30′, with single or multiple stems and graceful open fronds with rather soft foliage. Grows in subtropics. Many species also grow indoors.

For how to grow, see palm. Unlike most palms, however, this one should be planted in light shade in a place protected from wind. The smaller species do beautifully in wide, deep tubs.

CHAMAELAUCIUM Geraldton waxflower. Evergreen shrub to 12′ with needle-like leaves and sprays of waxy, pink or rose flowers in winter. Grows in warmest climates.

Plant in spring or fall in sun in average, well-drained, sandy or gravelly soil. Water in dry fall or winter weather but only sparingly during the summer. Fertilize in the fall. Prune hard after flowering or plant will become unkempt.

CHAMAEROPS HUMILIS Small palm tree to 10′ with clustered stems and fan-shaped green or silvery fronds. Subtropical.

For how to grow, see palm. Plants start slowly.

CHAYOTE *Sechium edule*, Christophine chuchu, vegetable pear, mirliton. Unkempt, tuberous-rooted vine to 50′ with edible, squash-like fruits in the fall. Tubers and foliage are also edible. Grows in warm climates.

Plant the whole fruits in rich soil 10′ apart in early spring. The stem end should be partially exposed. To insure pollination, at least two plants must be grown. The vine needs sun. Water in dry weather. Fertilize in spring and again in summer. Provide a sturdy trellis. Prune hard in the fall. In cold climates chayote may be grown as an annual vine, but it won't bear fruit as a rule.

CHERRY *Prunus.* Deciduous trees to 80′ with white or pink spring flowers and small, sweet or sour fruits. Sweet cherries grow in mild climates. Sour cherries grow in mild and somewhat cooler climates.

To assure fruiting, plant two varieties of sweet cherries; sour cherries, however, are self-fruitful. Plant in early spring or fall in full sun in a high location. The soil should be a deep, well-drained, sandy, rich loam. Space trees 20′–30′ apart and set them a little deeper than they previously grew. Water in dry weather until the end of August. Fertilize in early spring at the rate of 1 lb. of plant food per tree's year of age up to a maximum of 15 lb. (Sour cherries need about a third less than this.)

Prune cherries in late winter. Train young trees to 3–4 main branches (as with apples); and when fruit production starts, keep the center of the tree rather open so that air and light can enter. If fruit does not seem large enough but is plentiful, thin it out in the summer. As a general rule, sour cherries need more pruning than sweet.

To control insects and diseases, follow this schedule: In late winter when buds swell, apply dormant oil spray. Then apply all-purpose fruit tree spray immediately before blossoms open, at time of petal fall, two weeks later, two weeks after that, two weeks before fruit is ready to pick and right after fruit is picked. Do not leave fruit hanging on trees. Burn fallen fruits and leaves.

CHERVIL *Anthriscus cerefolium.* Annual herb to 2′ with parsley-like foliage used for flavoring. Grows anywhere.

Sow seeds in rows 18″ apart in early spring, and thin plants to 8″ Grow in light shade in average, well-drained soil. Keep weeded and watered. To dry, pick stems before plants flower and spread out in an airy, shady place on wire netting. Store in tight jars.

CHESTNUT *Castanea.* Deciduous trees to 90′ with familiar nuts in prickly burs. Grows in temperate climates.

The old American chestnut has been almost entirely wiped out by the chestnut blight fungous disease. The species now available are oriental and European trees and newly developed hybrids. None is as good as the American chestnut. If you insist on planting chestnuts, ask your state agricultural experiment station which trees they recommend for your area.

Plant in spring or fall in sun in average, well-drained, sandy soil. Water in dry weather. Fertilize every 2–3 years.

CHICK PEA *Cicer arietinum,* garbanzo. Bushy, 2′ annual producing small pods filled with seeds that are pointed like a ram's horn. These are roasted, used in soup or ground into meal. Grows in southwest.

In the spring, when weather is warm, sow seeds 2″ deep in rows 2′ apart in average soil in a sunny location. Thin plants to 5″. Keep weeded. Water in very dry weather, but the chick pea needs less moisture than most plants and does not need fertilizer unless the soil is very poor. Let peas mature before harvesting.

CHICORY *Cichorium intybus,* succory, witloof, French endive. Perennial to 6′ with fleshy roots which may be ground up and used instead of coffee. But main use of plant is in providing tightly bunched, slender white leaves used in salads. Grows in most parts of the country.

Sow seeds ½″ deep in average soil about June 1. Rows should be 2′ apart; seedlings thinned to 6″–8″. Keep soil damp throughout growing season. In the fall, before the ground freezes, carefully dig up the roots (which should be ¾″–1″ in diameter) and cut off the tops. Pack the roots upright and close together in a deep box. Fill in around them with sand or sandy soil, and cover with 6″ of sand. Water well and set in a basement at 55°. The roots will send up a tight head of leaves. As soon

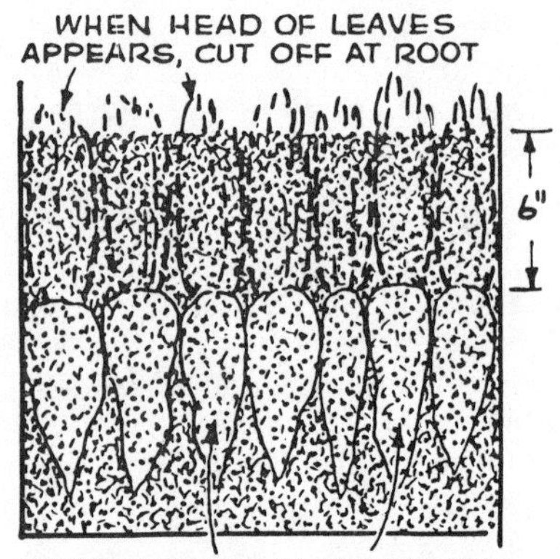

CHICORY ROOTS IN SANDY SOIL

as the tips of these appear at the surface, reach down through the soil and cut them off just at the roots.

CHILEAN GUAVA *Myrtus ugni.* Evergreen shrub to 6′ with small shiny leaves, pinkish spring flowers and purple-red, edible fruits. Grows in warmest climates.

Grow like myrtle (which see).

CHINABERRY *Melia azedarach,* China tree, pride of India, bead tree. One variety known as Texas umbrella tree. Wide-spreading, deciduous tree to 50′ with fragrant, lilac flowers in spring. Grows in warm climates.

Plant in spring or fall in sun in average soil. Water in dry spells. Fertilize every 2–3 years. This is a fast-growing tree that does well with little attention.

CHIN CACTUS *Gymnocalycium.* Round desert cacti to 8″ with step-like protuberances on the ribs. Grows in hot dry climates; also indoors.

For how to grow, see cactus, desert. But note that this plant prefers a soil containing somewhat more loam than is used for most cacti.

CHINESE CABBAGE *Brassica pekinensis*, celery cabbage. Leafy vegetable to 20″ resembling Cos lettuce. Does best in cool climates.

Sow seeds outdoors three months before first fall frost. Grow in average soil in which general-purpose fertilizer rich in nitrogen has been mixed. Grow in rows 18″–24″ apart and thin plants to about 10″. Keep well watered and cultivate often. Chinese cabbage needs full sun, but will bolt if weather becomes too hot.

CHINESE EVERGREEN *Aglaonema simplex*, Chinese water plant. Plant to 3′ with long leaves. Grows indoors.

Propagate by stem cuttings in water at any time. Plant in general-purpose potting soil. Keep moist and spray foliage with water occasionally. Feed every 2–3 months. Grow in a north window.

CHINESE FORGET-ME-NOT *Cynoglossum*, hound's tongue. Biennial to 2′ with sprays of forget-me-not-like flowers, mostly blue. Grown as an annual almost everywhere.

Sow seeds indoors 8–10 weeks before last frost. Transplant into individual pots. Then, when weather is safely warm, plant in the garden in average soil in full sun or partial shade. Fertilize when established. Keep watered.

CHINESE FOUNTAIN PALM *Livistona chinensis*, Chinese fan palm. A shrub-like palm to 80′ with numerous stems and large fan-shaped fronds. Grows in subtropics; also indoors.

For how to grow, see palm.

CHINESE HAT PLANT *Holmskioldia sanguinea*. Evergreen shrub to 10′ with somewhat straggling branches and red flowers. Grows in warmest climates.

Propagate by layering. Plant in spring in average soil in sun or light shade. Water in dry weather. Fertilize in spring. Prune after flowering.

CHINESE HIBISCUS *Hibiscus rosa-sinensis*, rose of China, China rose. Shrub to 20′ with beautiful, large flowers in many colors. Grows in warmest climates; also indoors.

Outdoors. Propagate by stem cuttings of half-hard wood in summer. Plant in spring in full sun in good, well-drained soil. Fertilize every month from February through August. Use general-purpose plant food and give a cupful at a time. If planted in limestone soil, spray with a nutritional spray in March. Water thoroughly once a week in dry weather. Prune lightly after flowering to control shape and size. Spray with malathion to control white snow scale. Extra-strong oil emulsion spray may also be used but should be rinsed off a half hour after it is applied. Regular spraying with malathion also controls thrips.

Indoors. Pot in general-purpose potting soil in the spring. Keep watered and fertilize monthly. Grow in a warm south window. Move outdoors in summer. In fall and winter, while plant is resting, reduce water supply considerably. Prune in spring before repotting and in fall when you bring plant indoors. You may have to cut it back very hard to keep it from overwhelming you.

CHINESE LANTERN *Physalis alkekengi.* Perennial often grown as an annual; to 2′, with delightful red, lantern-shaped fruits. Grows almost everywhere.

Sow seeds indoors 8–10 weeks before last frost. Shift seedlings into individual pots, then transplant outdoors when weather is warm. Grow in sun in good soil enriched a couple of times during the growing season with general-purpose fertilizer. Keep soil moist. Spray with malathion to control flea beetles. To dry the lanterns, place stems horizontally on chicken wire in a shady place.

CHINESE PARASOL TREE *Firmiana simplex*, phoenix tree, Japanese varnish tree. Deciduous tree to 45′ with handsome leaves, green bark, greenish flowers and strange fruits. Grows in warmest climates.

Plant in spring or fall in average soil in sun. Water in dry weather. Fertilize every 3–4 years.

CHINESE PISTACHE *Pistacia chinensis.* Spreading deciduous tree to 60′ with walnut-like foliage which turns brilliant red in fall. Grows in warm climates. Likes long hot summers especially.

Plant in spring or fall in full sun in average soil. Keep young specimens staked for several years. Water in dry weather. Fertilize in spring, especially when plant is young.

CHINESE WINDMILL PALM *Trachycarpus fortunei.* Tree with fan-shaped fronds atop a sturdy 40′ trunk which has a tough, dark, fibrous covering. Can survive 5° temperatures along seacoasts.

For how to grow, see palm. This is a slow-growing, undemanding tree that tolerates considerable neglect.

CHINESE WINGNUT *Pterocarya stenoptera.* Shallow-rooted, deciduous tree to 60′ with drooping flowers followed by long, drooping clusters of winged nuts. Grows in warm and mild climates.

Plant in spring or fall in sun in almost any soil. Water in dry weather. Tree is about as easy going and undemanding as ailanthus.

CHINQUAPIN *Castanea pumila.* Shrubby tree to 20′ with yellow-green leaves and small edible nuts. Grows in warm and mild climates from Texas east. See also golden chinquapin.

Grow like chestnut (which see).

CHIVES *Allium schoenoprasum.* Eight-inch perennial member of the onion family, used for seasoning. Grows in cool climates.

Divide old plants into sections each containing several bulbs and plant about ¼″ deep and 18″ apart in any good garden soil in the spring. Give plenty of sun. Keep watered and weeded. Sidedress lightly once during the summer with fertilizer. In the fall, before frost, lift a few clumps, set in large pots and grow in a cool south window. Plants left outdoors will continue to multiply and should be divided every 2–3 years.

CHOKEBERRY *Aronia.* Deciduous shrubs to 8′ with white spring flowers and bitter red or black fruits in fall. Grows widely east of the Mississippi.

Plant in spring or fall in sun or

light shade in average soil. Water in dry weather and fertilize every other year for best results. Propagate by layering.

CHORIZEMA VARIUM Bush flame pea. Evergreen shrubs to 2′ with leaves like the holly and a profusion of sweet-pea-like, reddish flowers in winter and spring. Grows in subtropics.

Propagate by stem cuttings after flowering. Plant in fall in sun or partial shade in average, well-drained soil. Water in dry spells. Fertilize in fall. Prune hard after flowering.

CHRISTMAS CACTUS *Zygocactus truncatus*, crab cactus. Spineless jungle cactus with many pendulous branches about 2′ long and red flowers about Christmas time. A house plant.

Pot in equal parts of loam, humus and sand. Be sure there is a good layer of coarse drainage material in bottom of pot. Grow in a warm east or west window away from drafts and hot radiators. About November, when plant starts making growth, water thoroughly whenever soil feels dry. Continue until after flowering, then water sparingly. Keep dry during October. Fertilize with liquid plant food about November 1 and every three weeks thereafter until flowering starts. During summer, plunge pot in the garden in light shade. Repot every two years. Propagate by stem cuttings planted shallowly in damp sand in November.

CHRYSALIDOCARPUS LUTESCENS Areca palm. Tree to 15′ with several yellow-ringed trunks, arching, feathery fronds and purple-black fruits. Grows in warmest climates; also indoors.

For how to grow, see palm.

CHRYSANTHEMUM Popular fall-blooming perennial to 40″ with flowers of various sizes and shapes and in almost all colors. Grows almost everywhere, but the hybrid varieties now sold are frequently killed by cold weather (though this isn't admitted always by the growers). Also a house plant.

Outdoors. Chrysanthemums are most easily propagated in the spring by prying small rooted sections with two or three shoots from big clumps. Planting can be done any time up to July 1. In milder climates, fall planting is also possible. Plant in well-drained, moisture-retentive soil in a place where plants will get a full day of sun. Dig soil to a depth of about 15″ and mix in peatmoss and a half cupful of general-purpose fertilizer per plant. After plant is in the ground, water thoroughly and do not let the soil dry out thereafter. A mulch of peatmoss will hold down weeds and hold in moisture. Avoid getting water on the foliage.

To produce stocky, well-branched plants with a lot of big flowers, pinch the stem tips when stems are 6″ long. Then pinch the tips of the new growth that comes when it is 6″ long. Don't pinch back growth after August 1, however.

Apply another half cupful of fertilizer when plants are 1′ tall. Stake tall varieties at that time. Spray or dust weekly with malathion and zineb. Burn diseased leaves and buds.

Protect flowers of later blooming varieties from early frost by covering at night with a tent of burlap, paper or plastic. When plants die down in late fall, cut stems to the ground and cover with leaves, salt hay or evergreen boughs. In very cold climates new varieties of chrysanthemums should be moved into a coldframe.

Well-grown plants usually need to be divided every two years and perhaps even every year.

One of the unusual things about chrysanthemums is that they can be transplanted successfully even after they have started blooming. This means that it is possible to fill your flower beds in the fall with blooming plants purchased from a grower or moved from another part of your garden. Always keep these plants watered well. Unfortunately, however, you can't count on their living through the winter in cold climates.

Indoors. To make hardy garden chrysanthemums bloom indoors in the fall, keep outdoor plants pinched back during the summer. Then in September, when buds are developing, water well and shift plants into large pots in general-purpose potting soil. Move into a very cool (under 50°), sunny window. Spray foliage daily and keep soil evenly moist. When plants are through flowering, either throw them away or cut them back to the ground and move them into a coldframe. Take cuttings from them in the spring.

Florist chysanthemums should be kept in a cool, sunny window, watered well and sprayed daily with water. In the north, only hardy chrysanthemums can be saved to grow outdoors the next year. In the south, the big greenhouse types may also be saved.

CINERARIA *Senecio cineraria.* Foot-tall florist's plants with handsome foliage and flower masses in all but yellow. A temporary house plant. Also grows outdoors in California.

Indoors. This plant is too difficult to raise yourself from seeds. When you buy or are given a plant, keep it in a south window at no more than 60°. Water regularly.

Outdoors. Buy plants. Plant in fall in well-drained, deeply dug soil containing considerable humus and some dried manure. Grow in partial shade in wind-free areas. Keep soil moist and spray leaves with water on warm days. Fertilize monthly with liquid plant food. Protect from snails. Spray with malathion to control aphids and leaf miners.

CLARKIA Annual to 3′ with lovely flowers in white, rose, pink and purple. Grows best on the Pacific Coast and in regions with cool summers.

Sow seeds where they are to grow in early spring. Grow in average, well-drained soil in partial shade. Thin to stand 9″ apart, and fertilize at that time. Water in dry weather.

CLAYTONIA Spring beauty, mayflower, grass flower. Plant growing from a corm to 6″, with lax clusters of white or pink-tinged flowers in very early spring. One or more species grows in the different parts of the country.

Plant corms 3″ deep and about 3″ apart in moist, average soil in shade in the fall.

CLEMATIS A large group of gorgeous vines to 20′ with deciduous or evergreen foliage and flowers in many colors. Grows almost everywhere if you pick the right species.

Clematis can be propagated by layering in the spring, but you'll do best to start with two-year-old plants grown by a professional. Be sure to find out and write down the peculiar habits of the plants you buy, because clematis is not completely uniform in its needs or the way you handle it.

Plant clematis in a spot where it will get at least six hours' sun a day. The roots, however, should be kept shaded either by a wall or mulch. Don't set the plants in the vicinity of cedars or arborvitaes. Allow 3′ of elbow room between the vines and other large plants.

Planting is best done in the spring in cold climates; in spring or fall elsewhere.

Generally the vines prefer a sweet (limed), rich soil. Dig a hole 2′ wide and 2′ deep and pour 2″–4″ of coarse gravel in the bottom to provide excellent drainage. Mix soil with humus and a small amount of general-purpose fertilizer. If soil is very heavy, add some sand.

Set the new plant in the soil so that the crown is 2″ below the surface. Fill in lightly around the roots and water well. Tie to a stake. Then mulch the soil around the plant with a thick layer of peatmoss. In the summer or fall, fertilize lightly.

Routine care of established plants is as follows: Fertilize in spring and again after flowering. Apply lime in spring if your soil needs it. Cultivate sparingly and shallowly, but maintain a thick mulch at all times to keep down weeds and hold in moisture. Train the stems on strings, wire or a light trellis. In winter, in cold climates, mound soil up around the plants. Don't prune too violently. As a rule, it is necessary to remove only the dead, weak or damaged stems.

CLERODENDRON Different species called glory bower; tubeflower, skyrocket plant and Christmas star. Evergreen vines to 15′ and shrubs to 9′ with long-lasting, red and white flowers. Grows in subtropics; also a house plant.

Outdoors. Propagate by seeds or stem cuttings. Plant in good soil to which plenty of humus and some fertilizer have been added. Grow in light shade or sun. Keep well watered. Fertilize annually in early spring and again in summer. Mix new humus into the soil in the fall. Spray with malathion to control insects. The vine should be trained and tied to a sturdy trellis. Remove flowers when they die. Cut back old wood in winter (flowers come on new wood).

Indoors. Pot in general-purpose potting soil with extra humus. Place in a south window. Spray frequently with water. Keep soil moist except in fall and winter, when watering should be reduced. Other requirements as above.

CLEYERA *Eurya.* Shrubs to 10′ with glistening evergreen foliage and red berries. Grows in warm climates.

Should be treated like *camellia* (which see). Propagate by softwood stem cuttings. Grow in sun or partial shade. Don't expose to wind.

CLIFF BRAKE *Pellaea atropurpurea.* Leathery fern to 2′ with purple stems. Grows mainly in the east.

For how to grow, see fern. Cliff brake does well in ordinary, dry soil and in sun or partial shade.

CLIMBING FERN *Lygodium palmatum,* Hartford fern. Climbing, evergreen fern to 4′ with dainty fronds. Grows in all but cold and very dry climates.

Propagate by division of roots. Grow in shade, out of the wind, in light, acid soil with a large amount of humus. Keep moist. Do not fertilize. Provide strings or wires for twining stems to climb on. Mulch with leaves in the fall. Remove dead fronds promptly.

CLIMBING HYDRANGEA *Hydrangea petiolaris.* Deciduous vine to 50′ with large, white flower clusters in early summer. Grows in all but most severe climates. See also *hydrangea.*

Propagate by layering. Plant in good, well-drained soil and keep well watered for the first several years. Thereafter, water in dry weather. Fertilize in early spring. Grow in sun or partial shade. Prune to control size and improve appearance in early spring. Provide a rough surface for the aerial rootlets to cling to.

CLIMBING YLANG-YLANG *Artabotrys odoratissimus.* Evergreen vine to 10′ with very fragrant, yellow-green flowers and grape-like bunches of inedible fruits. Grows in subtropics.

Soak seeds for 12 hours, then sow in pots in winter. Shift plants in spring to rich, humusy, well-drained soil in sun or partial shade. The vine needs a very sheltered location—under a porch eaves, for example—to survive frost. Water in dry weather. Fertilize in winter and again in spring. Prune after fruiting to control growth, which is moderate. Provide a trellis or some other support for tendrils to cling to.

CLINTONIA BOREALIS Blue beads, cow tongue. Perennial wildflower, 15″ tall, with ladyslipper-like leaves and bell-shaped, greenish-yellow flowers and blue berries. Grows in cold climates.

Dig up in late summer, being sure to take at least one of the underground runners from which new plants will develop. Plant in shade in cool, moist, humusy soil.

CLIVIA Evergreen perennial to 16″ with magnificent clusters of lily-like, orange flowers in spring. Grows indoors; also outdoors in warmest climates.

Indoors. Pot the fleshy roots in general-purpose potting soil so that the crown is at soil level. Over summer, keep outdoors in partial shade, water when soil feels dry and fertilize every fortnight. In the fall, bring the plant indoors into a light, cool (under 55°) place and keep it there until January. Water sparingly and do not fertilize. Then shift into a warm, sunny window, keep soil moist and feed every fortnight. Reduce watering slightly after flowering ends. Don't repot clivias more than once every three years. Divide the roots only when they have outgrown every container you own.

Outdoors. Handle pretty much as above. You are probably better off growing the plant in pots, but you can set it directly in the garden if you are sure it won't get hit by frost or drowned by winter rains. Grow in light shade. Fertilize in late winter, late spring and fall.

COB CACTUS *Lobivia.* One species known as golden Easter lily cactus. Desert cacti to 1′ with round, ribbed, spiny bodies and lovely lily-like flowers in red, yellow and white. Grows in warmest climates; also indoors.

For how to grow, see cactus, desert.

COBRA PLANT *Darlingtonia californica*, California pitcher plant. An insectivorous plant with 30″ leaves shaped like a cobra dancing to music. Grows on the Pacific Coast; also indoors.

Grow like pitcher plant (which see).

COCCOTHRINAX Palm trees to 30′ with single trunks and fan-shaped leaves that are green on top and silvery beneath. Subtropical.

For how to grow, see palm. This genus needs full sun and can be exposed to it when quite young.

COCCULUS LAURIFOLIUS Evergreen shrub to 25′ with large, leathery, glossy leaves on arching stems. Grows in warm climates.

Propagate by half-ripe stem cuttings in summer Plant in spring or fall in shade in average soil. Water in dry spells. Prune in fall to control shape and discourage legginess.

COCKSCOMB *Celosia.* Annuals to 3′ with plumed or strangely shaped, crested flowers mainly red but sometimes yellow. Grows almost everywhere.

Sow seeds indoors about eight weeks before last frost is expected in spring, Shift into a sunny location in average soil. Water in dry weather. Fertilize when plants are established.

COCONUT PALM *Cocus nucifera.* Towering (to 100′), usually leaning palm with a topknot of fronds and the familiar hard-shelled fruit. Grows in subtropics.

For how to grow, see palm. Coconut palms are not injured by salt spray and may therefore be grown close to the sea.

COCOZELLE *Cucurbita.* A long, cylindrical, dark-green summer squash. Grows almost everywhere.

Grow like summer squash (see squash).

COELOGYNE CRISTATA Evergreen, epiphytic orchid to 1′ with drooping clusters of white flowers in winter. A house plant.

For how to grow, see orchid.

COLCHICUM Autumn crocus, meadow saffron. Bulbous plants to 1′ with large, crocus-like flowers in various colors in the fall and with foliage the next spring. Grows best in cool climates.

Plant bulbs 3″–4″ deep and 6″ apart in August as soon as they are available. Grow in average, well-drained soil. Water in dry weather. Fertilize lightly in early spring. Allow foliage to mature in the summer no matter how unsightly it is.

COLEUS Painted nettles. Tender plant to 3′ grown for its colorful leaves. A house plant; also grows outdoors.

Indoors. Propagate from seeds or by stem cuttings rooted in water. Pot in general-purpose potting soil. Water regularly. Fertilize every month. Grow in a sunny window. Keep plant pinched to prevent legginess. Take cuttings in spring and discard the rest of the plant. Spray with water to eliminate mealybugs.

Outdoors. Grow from cuttings taken from house plants, or sow seeds indoors 8–10 weeks before last frost. Shift directly into humusy garden soil or into pots. Grow in sun or light shade. Water, feed and pinch as above. Take cuttings in August or September for development as house plants.

COLLARDS *Brassica oleracea acephala.* A 3′ variety of kale grown mainly in the south for greens.

Collards need a well-drained, limed, humusy soil. Fertilize with nitrogen-rich general-purpose plant food at planting time. Sow seeds in rows 2′–3′ apart as soon as soil can be worked. Thin plants to 18″–24″. Keep watered and weeded.

COLLINSIA BICOLOR Chinese houses. Annual to 2′ with white and rose, snapdragon-like flowers arranged in tiers. Grows mainly in California.

Sow seeds where they are to grow in early spring (or in fall in areas with mild winters). Grow in partial shade in rich, humusy soil and fertilize when plants are thinned to 9″–12″ apart. Feed again about six weeks later. Water regularly.

COLUMBINE *Aquilegia.* Perennial to 3′ with enchanting, strangely shaped and sometimes strangely colored flowers in spring and summer. Grows in all but hottest climates.

Sow seeds in spring for flowers the next year. Transplant in fall or spring. Divisions can be made in early spring. Grow in sun or partial shade. Plants do well in average, well-drained soil but better in well-drained, humusy soil. Fertilize in early spring and again about June. Water regularly.

COLUMNEA Plant to 15″ with tubular pink, red or yellow flowers. A house plant.

Propagate by greenwood stem cuttings. Pot in equal parts of sterilized loam, humus and sand. Grow in a warm east or west window. Keep soil moist. Spray foliage daily with water. Fertilize every 3–4 weeks. Pinch young growing stems to promote bushiness.

COLUTEA Bladder senna. Deciduous shrubs to 4′ with yellow or brownish-red, summer flowers followed by bladder-like, red pods. Grows in mild and warm climates.

Sow seeds in spring after the ground has warmed up. Transplant in fall or spring. Grow in a sunny spot in average soil that is well drained. Water sparingly and feed only every 2–3 years.

CONEFLOWER *Echinacea,* purple daisy, black Sampson. Coarse perennials to 3′ with rose or purple ray flowers. Grows in all but extreme climates.

Divide roots every third year in the spring and plant in good, sandy soil in sun or partial shade. Water only in dry weather. Fertilize lightly in spring. Mulch in winter in cold climates.

CONGEA TOMENTOSA Woolly congea. Evergreen vine to 40′ with long-lasting lavender bracts in winter. Grows in subtropics.

Plant in average soil in sun in spring or fall. Water in dry spells. Fertilize a couple of times when making growth. Prune after flowering.

CORAL PLANT *Russelia equisetiformis,* fountain plant, firecracker plant. Four-foot shrub with profuse red flowers. Grows in warmest climates.

Propagate by stem cuttings in spring. Plant in spring or fall in full sun in average soil. Water in dry weather. Fertilize every 2–3 years. The plant also does well in hanging baskets if it is watered almost daily and fed every month.

CORAL VINE *Antigonon leptopus,* corallita, rose de Montana, pink vine, confederate vine, mountain rose, love's chain, Queen Anne's wreath. Tuberous-rooted, deciduous vine to 40′ with masses of rose-pink flowers in summer and fall. Grows in subtropics.

Grow from seeds, stem cuttings in the spring or division of the tubers. Plant in deeply dug, well-drained soil of poor to average quality. Do not fertilize. Needs full sun but dislikes wind. Water heavily in spring and summer, then gradually withhold moisture. In winter, water just enough to keep tubers from shrivelling. Cut back hard in the fall after flowering.

COREOPSIS Perennial to 30″ with daisy-like flowers, usually yellow. Grows almost everywhere. See also calliopsis.

Sow seeds outdoors in a seedbed from early spring until July; or propagate by root division in early spring. Set into the garden in spring or fall. Grow in full sun in average soil. Water in dry weather. Fertilize in early spring.

CORIANDER *Coriandrum sativum.* Annual herb to 15″ with unpleasant smelling, feathery foliage, pink flowers and seeds used for flavoring. Grows almost everywhere.

Sow seeds after danger of frost is past in average, well-drained soil in full sun. Fertilize when thinning the young plants to about 9″ apart. Water in dry weather.

CORK TREE *Phellodendron.* Deciduous trees to 50′ with aromatic leaves, black fruit and corky bark. Grows almost everywhere, but does best where winters are fairly cold.

Plant in spring or fall in a sunny location. Soil should be deep and of average quality. Water in very dry weather. Fertilize for a few years after planting; thereafter only about every 3–5 years.

CORN *Zea mays.* Favorite vegetable which grows almost everywhere. Late varieties are up to 8½′; early varieties are smaller.

Corn is not very particular about soil (witness the fact that it is at its best in thin, rocky New England soils), but it appreciates the addition of fertilizer at planting time. Full sun is a necessity. Make the first sowing one week before last frost. For a summer-long succession of crops, plant early varieties at two-week intervals or you can make several at-the-same-time plantings of early, medium and late varieties.

Sow seeds 1″ deep in rows 30″ apart. Space seeds 3″ apart and thin to 6″. Cultivate regularly to keep free of weeds. Hill up soil 6″–8″ around base of stalks as they develop. Water in dry spells. Do not bother to pull off suckers that form at base of plants. Pick ears as soon as kernels are plump just before they are to be eaten.

To control borers, dust the tassels several times with Sevin. Also apply Sevin to young silks to stop earworms. When a black, powdery fungus called smut develops on the ears, burn ears and stalks promptly.

CORN SALAD *Valerianella locusta olitoria,* fetticus, lamb's lettuce. Annual to 1′ with tender, almost tasteless, rosette-shaped leaves that are used in salads or cooked like spinach. Grows almost everywhere, but only in cool weather.

Sow seeds in early spring in sun in average soil which has been lightly limed and fertilized. Space rows 1′ apart and thin plants to 4″–6″. Water in dry weather. Keep weeded.

CORNFLOWER *Centaurea*, bachelor's button, ragged robin, bluebottle. Annual to 2′ with small, ragged, blue flowers. There is also a perennial version. Grows almost everywhere.

Annual. Sow seeds where plants are to grow in very early spring or late fall. Give full sun, a well-drained, average soil. Fertilize when plants are thinned. Water in dry weather. Keep fading blooms picked off.

Perennial. Sow seeds in a seedbed any time until July. Transplant in fall or early spring. The plants have the same requirements as the annuals.

CORREA Evergreen shrubs to 30″ and spreading further, with dense foliage and bell-shaped pink, red or white flowers in winter. Grows mainly in California's warm areas.

Plant in partial shade in average soil in fall. Water until plants are established. Fertilize very occasionally. Otherwise, leave alone except for pruning after flowering.

CORYLOPSIS Winter hazel. Deciduous shrubs to 6′ with yellow flower clusters in late winter. Grows in mild and warm climates.

Propagate by layering. Plant in spring or fall in a sunny, sheltered place. The soil should be of average quality with extra sand added to ensure good drainage. Water in dry spells. Fertilize every couple of years.

COSMOS Annual to 4′ with pink, rose, white or yellow ray flowers. Grows almost everywhere.

Sow seeds indoors 6–8 weeks before last frost, or sow outdoors where plants are to grow when soil dries out and is warm. Grow in average, well-drained soil in sun or partial shade. Water in dry weather. Fertilize when young plants are established. Pinch tops when plants are 1′–2′ tall to promote branching. In windy locations staking is required.

COSTMARY *Chrysanthemum balsamita.* Perennial to 3′ grown for its fragrant leaves. Grows in temperate climates.

Divide plants in spring and plant in sun in average soil. Water in dry weather. Fertilize in spring.

COSTUS Spiral flag, spiral ginger. Evergreen perennials to 10′ with handsome, spirally arranged leaves and yellow, red or white flowers in summer and fall. Grows in warmest climates; also indoors, but only if it has plenty of room.

Propagate by division of the roots or by stem cuttings. Plant in spring in partial shade (or an east window) in the equivalent of equal parts of loam, humus and sand. Keep moist and fertilize several times while making growth. The plant needs plenty of warmth.

COTONEASTER Deciduous or evergreen shrubs to 12′, of various growth habits, and with white or pink spring flowers and red or black fruits. Grows in mild and warm climates.

Propagate by layering or by green stem cuttings over heat. Plant in spring in full sun in average, well-drained soil. Once planted, do not move. Water in dry spells. Fertilize every year or two. Prune in late winter when berries fall. Spray with miscible oil before growth starts if scale becomes a problem.

COTYLEDON Succulents to 4′ with beautiful thick leaves and large clusters of hanging flowers in various colors. Grows in warmest climates; also indoors. Some species withstand some frost.

For how to grow, see cactus, desert. In watering cotyledons, avoid wetting leaves since that spoils the powdery white "bloom."

CRAPE JASMINE *Ervatamia coronaria*, fleur d'amour, East Indian rosebay. Also identified as *Tabernaemontana coronaria.* Evergreen shrub to 6′ with glossy leaves and white summer flowers. Grows in subtropics.

Propagate by semi-hardwood cuttings. Plant in spring or early fall in full sun in good soil containing a fair amount of humus and sand. Water in dry weather. Fertilize in spring. Prune in winter.

CRAPE MYRTLE *Lagerstroemia.* Shrubs or trees to 20′ with lilac-like flower clusters in white, red or purple in summer. Grows in warm climates.

Propagate by stem cuttings in spring and set out the plants in the fall. Or plant young purchased specimens in fall. Grow in full sun in well-drained, fertile soil. Water deeply but not too often in dry weather. Fertilize lightly in spring. Spray with malathion to control aphids and with phaltan to prevent mildew. If left alone, crape myrtle develops into a large plant. However, you may want to cut the shoots back hard every winter.

CRASSULA Succulents to 10′ with pretty, fleshy leaves arranged in cross shape on the stems. One well-known species called jade plant. Others known as scarlet paint brush, string of buttons, silver dollar, St. Andrew's cross, elephant grass, silver beads. Grows in hot, dry climates; also indoors.

For how to grow, see cactus, desert. Make sure the soil is very well-drained and that plants grown indoors receive ample air as well as sun. If grown in pots, repotting once a year to the next size larger pot will probably be necessary.

CREEPING FIG *Ficus pumila.* Evergreen vine to 50′ with small, heart-shaped leaves that form a flat mat against the plant's support. Best known as a house plant, but also grows outdoors in subtropics.

Indoors. Propagate by stem cuttings. Pot in general-purpose potting soil which should be kept moist. Grow in an east or west window. Fertilize lightly every two months. Spray leaves with water now and then. The aerial rootlets will cling to any surface. Pinch back ends of stems occasionally to control growth and promote branching at the base.

Outdoors. Plant in very well-drained soil containing extra humus. Grow in partial shade; not on a hot wall facing the west or south. Keep watered and spray foliage with water often. Prune lightly in the fall. Control aphids by spraying with malathion.

CRENSHAW MELON *Cucumis melo.* Large melon with yellow skin and salmon-pink flesh. Strain which matures in 110 days is grown in warm climates. Fast-maturity strain can be grown in cooler climates.

CRESS *Lepidium sativum,* pepper grass. Annual herb to 1′ with pungent leaves used in salads. Grows best in cool climates. One type known as winter cress survives very severe cold. See also water cress.

Cress prefers good, well-drained soil that has been limed and improved with fertilizer. Sow seed in rows 1′ apart as soon as soil can be worked, and make several additional sowings up to May. Winter cress should be sown in very early spring or fall. Don't bother to thin plants. Cut with scissors whenever a supply is needed for salads.

CRINODENDRON DEPENDENS Lily-of-the-valley tree, white lily tree. Evergreen tree to 25′ with spreading limbs and a profusion of white, bell-shaped flowers in summer. Grows in mild and warm climates.

Plant in spring or fall in sun in average soil. Water often and deeply in dry weather and mulch the soil. Fertilize in spring. Prune in winter to develop tree shape; otherwise plant may become shrubby.

CRINUM Milk-and-wine lily. Bulbous plants to 4′ with many white or pink flowers in spring or summer. Grows in warm climates.

Plant in fall. Set bulbs just below the surface of rich, well-drained, humusy soil in a sunny location. Water regularly until flowering stops. Fertilize in early spring. Disturb as little as possible. Don't count on bloom for a couple of years after bulbs are planted.

CROCOSMIA AUREA Copper tip. Plant to 42″ growing from a corm with spikes of orange-yellow flowers in fall. Grows almost everywhere.

Grow like gladiolus (which see).

CROCUS Small (to 6″) bulbs that herald spring with their variously colored flowers. There are also autumn-flowering types. Grows best in cool climates. Also a house plant.

Spring-flowering crocus outdoors. Plant in sun in ordinary, well-drained soil in the fall. Bulbs should be 3″ deep, 4″ apart. To discourage rodents, soak bulbs in water, then roll in red lead powder.

Autumn-flowering crocus outdoors. Plant as above in late August or September.

Indoors. See How to Force Bulbs for Indoor Bloom. Use spring-flowering bulbs and don't try to force them too fast.

CROSSANDRA INFUNDIBULIFORMIS Evergreen shrub to 3′ with soft orange flowers much of the year. Subtropical.

Propagate by stem cuttings. Plant in spring or fall in partial shade in average soil. Water in dry weather. Fertilize in late winter and again in late spring. Prune in early winter. Spray with malathion to control mites.

CROTALARIA Rattle box, canary bird bush. Spreading shrubs to 12′ with clusters of yellow, sweet-pea-like flowers at different times depending on species. Grows in warmest climates.

Grow from spring-sown seeds

that have first been soaked in water for several hours. Set out plants in spring or fall in sun in average soil. Water in dry spells. Fertilize when starting into growth. Thin branches after flowering.

CROTON *Codiaeum.* Shrubs to 6′ with brilliantly colored leaves. Grows outdoors in subtropics. Also a house plant.

Outdoors. Propagate by air-layering when plants get leggy. Plant in spring or fall in partial shade in average soil. Water regularly. Fertilize once a year. Don't expose to subfreezing temperatures. Pick off mealybugs with cotton swabs dipped in alchohol. Spray with miticide to control red spider.

Indoors. Pot in general-purpose potting soil which should be kept moist. Grow in a warm south window which is lightly screened at midday. Spray plants with water frequently and keep pots on wet pebbles. Move outdoors into light shade in summer.

CROWBERRY *Empetrum nigrum.* Spreading, evergreen shrub to 10″ with purple flowers and black berries. Grows in cold climates.

Plant in spring in the sun. The soil must be well-drained but moisture retentive—with considerable sand and peatmoss. Mulch with peatmoss in summer to keep soil moist and cool.

CROWN CACTUS *Rebutia.* Round desert cacti only a few inches in diameter, producing many red or orange flowers in a circle at the base of the plant. Grows in hot dry climates; also indoors.

For how to grow, see cactus, desert. Grow in light shade outdoors. Indoors, plants need protection from hot summer sun only. Water regularly when making growth.

CROWN OF THORNS *Euphorbia splendens.* Creeping succulent with spiny 3′ stems and bright-red bracts in winter and often at other times of the year. A house plant.

Propagate in spring by stem cuttings which are allowed to dry for several days and are then planted shallowly in sand which has just a trace of moisture. Keep in a dim place until roots form. Then pot in equal parts of loam, humus and sand in a pot with a thick layer of drainage material. Grow in a south window. Water when soil feels dry but keep on the dry side in November and December. Fertilize with dilute liquid plant food every six weeks when plant is growing. Train on a small trellis if you wish. Put outside in summer.

CRYPTANTHUS Terrestrial bromeliads, rarely much more than 15″ high, shaped like a flat star, with usually white flowers. Grows outdoors in subtropics; also indoors.

Propagate by the tiny plants appearing between the leaves. Pot these shallowly in a mixture of one part sand and one part osmunda. Water, preferably with rain water, when soil feels dry. Give a little dilute liquid plant food at six-week intervals when plants are making growth and blooming. Grow in an east or west window or in light shade outdoors. Give normal house temperatures indoors; protect from frost outdoors. Spray foliage of indoor plants with water in winter. Spray with malathion to control scale insects, but rinse with water the next day.

CRYPTOMERIA Evergreen tree to 25′ with various shrubby varieties. The plants are dense, have needles with curved tips. Grows in temperate climates near the sea coast.

Plant in spring or early fall in average, well-drained soil. Give full sun. Water in dry weather. Fertilize every year or two when plant is young. Don't worry about the bronze coloration in winter.

CTENANTHE Perennials to 3′ with attractive foliage that is sometimes variegated. A house plant.

Pot in general-purpose potting soil. Grow in a warm north window. Keep watered and provide as much himidity as possible. Fertilize lightly with liquid plant food every two months.

CUCUMBER *Cucumis sativus.* Familiar vine vegetable. Grows anywhere.

Cucumbers prefer light, humusy soil but are not overly particular. Add well-rotted or dried manure if available; otherwise, mix in some general-purpose fertilizer at planting time. Sun and heat are essential.

Sow seeds after all danger of frost is past and ground has warmed up. Scatter 6–10 seeds in a circle about 1′ in diameter and cover with 1″ of soil. Each of these circles is called a "hill" (though it does not have to be mounded up). The hills should be 4′–5′ apart in all directions. When plants appear, thin out all but three. Cultivate lightly and keep out all weeds when the plants are young; if you don't you'll have a weed patch when plants are fully grown. Water thoroughly if you don't have rain once a week. In a very small garden where space is at a premium, cucumbers can be trained on a wire trellis.

Dust plants regularly with rotenone to control beetles and the diseases they spread. Burn all plant debris in the fall.

Note that planting cucumbers next to melons does not affect the flavor of the melons.

CUMIN *Cuminum cyminum.* Annual herb to 6″ with delicate foliage, white flowers and aromatic seeds. Grows almost everywhere.

Sow seeds in spring after danger of frost is past. Grow in sun in average, well-drained soil. Thin plants to 4″ and fertilize at that time. Water in dry weather.

CUNNINGHAMIA LANCEOLATA China fir. A conifer to 75′ with branches drooping at the tips and closely spaced, tapering leaves. Grows in warm climates.

Plant in fall or spring in partial shade in soil containing a fair amount of sand and humus. Water in dry spells.

CUP AND SAUCER VINE *Cobaea scandens*, cathedral bells, Mexican ivy. Dense woody vine to 25′ with bell-shaped, violet flowers in summer. Grown as a perennial in the south, as an annual further north.

Sow seeds indoors in winter. The seeds should be set on edge and barely covered with soil. Plant outdoors after danger of frost is past in well-drained, average soil. Keep moist. Fertilize at planting-out time and once or twice again during the summer. Grow in full sun in a warm place. Provide a sturdy, rough support to which tendrils can cling. In the fall, pull up plants grown as annuals or cut back perennials.

CUPHEA One species called Florida heather; another, cigar flower. Plants to 2′ with tubular, red, magenta or white flowers. Grows indoors; also outdoors in warmest climates.

Indoors. Propagate by stem cuttings and pot in a small pot of general-purpose soil. Grow in a south window. Water when soil feels dry. Fertilize in winter and spring at six-week intervals.

Outdoors. Grow in partial shade. Treat as above.

CURRANT *Ribes.* Deciduous shrubs to 6′ with juicy, edible, red, black or purple berries. Grows in cool and cold, moist climates.

Note: Do not plant currants if white pines are growing anywhere in the area, because the pines will be killed by white pine blister rust, for which currants are an alternate host. Some states prohibit planting of currants.

Plant in fall or very early spring in well-drained but fairly heavy soil. Grow in the sun in cooler climates; in light shade in somewhat warmer areas. Space plants 3′–5′ apart in rows 5′ apart. Water in dry weather. Fertilize in early spring. Keep soil cultivated shallowly, or mulch with leaves or straw. Prune in the fall to about nine main stems, none of which should be more than three years old (in other words, cut out all stems four years old or older because they do not bear much fruit). You should have some one-year-old stems, some two-year-old stems and some three-year-old stems. Spray with rotenone or nicotine sulfate to control aphids.

CYANOTIS One species called teddy bear plant, another, pussy ears. Trailing plants with stems a couple of feet long, hairy leaves and blue or rose flowers. A house plant.

Pot in general-purpose potting soil. Grow in an east or west window. When plant is making growth, water when soil feels dry; but reduce the supply during the winter rest period. Fertilize 2–3 times during spring and summer.

CYCLAMEN Tuberous plant with rounded leaves and red, pink or white flowers on 8″ stems. Familiar house plant. Hardy species grow outdoors in most climates.

Indoors. Cyclamens can be raised from seed but it is a long process. Buy plants. Grow in an east window at about 60°. Keep soil moist but not soggy. From start of bud formation to end of flowering, feed every 3–4 weeks. Spray with malathion to control mites if growth becomes weak and twisted. Don't try to save tubers.

Outdoors. In late summer plant tubers 2″ deep in well-drained, humusy soil. Grow in light shade. Fertilize sparingly with bone meal. Cover with leaves in winter.

CYMBIDIUM Evergreen, terrestrial orchids to 4′ with long spikes or sprays of flowers in whites and purples. A house plant.

For how to grow, see orchid. Pot in equal parts of osmunda, peatmoss and leafmold. Do not repot more often than every 3–4 years.

CYPRESS *Cupressus.* Distinctive, long-lived evergreen trees to 80′, usually spreading but sometimes narrow and erect, with scale-like leaves. Grows in warm climates.

Plant in early spring or fall in sun in well-drained average soil. Water deeply in dry spells. Fertilize occasionally when tree is young.

CYPRIPEDIUM Terrestrial orchids to about 18″ with flowers at different times of year in unusual shades of purple, green, brown, yellow, pink, etc. Some species are house plants. Others, commonly referred to as ladyslippers, are hardy and grow in various climates outdoors.

For how to grow hardy types, see ladyslipper.

For how to grow house plants, see orchid. Pot or repot after flowering in equal parts of osmunda, leafmold and peatmoss. The base of the plant should be slightly above the pot rim so that water will not stand in it. During spring and summer, keep shaded by cheesecloth for the better part of the day. Night temperature should not go below 70°. Spray with malathion if thrips cause streaked flowers, discolored foliage.

CYRILLA RACEMIFLORA Titi, leatherwood. Shrub to 20′, which may or may not lose its leaves, depending on the climate. Has white spring flowers. Grows in the south.

Plant in spring or fall in partial shade in average soil that is somewhat acid. Water in dry weather. Fertilize every 2–3 years. Prune hard after flowering. Pruning and ample moisture help to hold plant down in size.

DAFFODIL *Narcissus.* Favorite spring bulbs to 18″. Grows anywhere; also indoors. See also *narcissus.*

In beds. Daffodils grow in full or filtered sun in average soil that is neither too wet nor too dry. Dig soil to a depth of 1′, remove stones and mix in a little general-purpose fertilizer. Add humus if soil is very sandy. Plant bulbs in the fall 6″ deep and 6″ apart. Use a trowel to make the holes, not a dibble. Water well. No additional moisture is needed except in droughts. In cold climates, cover the bulbs the first winter with leaves, salt hay or evergreen boughs after the ground has frozen. In early spring, before growth starts, scatter fertilizer on the soil. After blooming, remove old flowers but don't touch the leaves until they are dry. If you dislike their appearance, bunch them together, fold down the top 5″ and secure with a rubber band. Since daffodil bulbs multiply rapidly, it is necessary to dig them up every 3–5 years and separate them.

In naturalized plantings. To plant daffodils informally in the lawn, fields or woods, you can make individual holes with a trowel. A faster way, however, is to use a pick. Ram the point into the soil, rock it back and forth to open up the hole, set in the bulb, making sure that it rests flat on the soil, not over an air space; then fill in hole. From here on, handle as above.

Indoors. See How to Force Bulbs for Indoor Bloom.

DAHLIA Tuberous plants to 8′ with beautiful flowers in many sizes and colors. Grows almost everywhere, but best treated in the deep south as an annual.

Dahlias, especially miniature types, are easily grown from seeds sown indoors 8–10 weeks before last frost and planted outside where they are to grow when weather is warm. All types of dahlias may also be propagated by division of the tubers. To do this, cut the clump with a sharp knife so that each division includes a piece of the old stem and a visible bud.

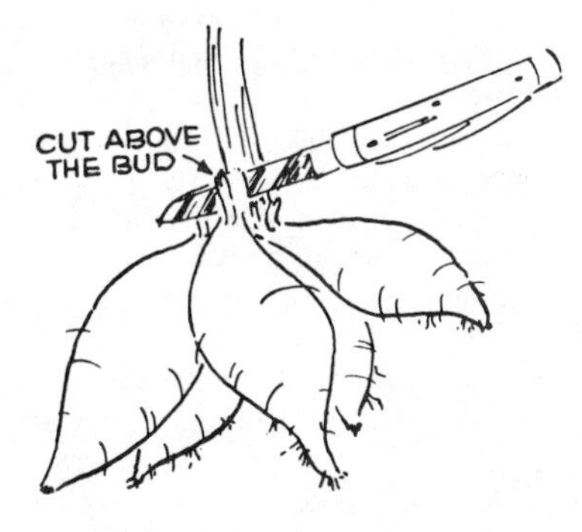

Plant the tubers outside when the weather is reliably warm. Dahlias need full sun and good air circulation. The soil should be well-drained, sandy loam enriched with wood ashes and a handful of bone meal or superphosphate.

For medium-size and giant dahlias, dig holes 3′ apart and 6″ deep and set stout stakes at the back of the holes. For small dahlias, space the holes 18″ apart, dig 6″ deep and forget the stakes. Lay the tubers flat in the holes with the buds pointed toward the stakes, and cover with 2″ of soil. Water well. As the plants grow, fill in the holes the rest of the way. Start tying plants to the stakes when they are 1′ tall. Pinch out the growing tips to promote bushiness. For largest flowers, remove all but one stem when plants are about 9″ tall. Nipping off the outer two flower buds in each group of three also makes for larger flowers.

Before they bloom, dahlias need to be watered only in very dry weather; but after they start flowering, soak the soil thoroughly every week. Keep cultivated. Apply general-purpose fertilizer with a low nitrogen content twice after the plants are well up. Spray plants weekly with malathion. Burn stunted plants that have dwarfed leaves with yellow spots or rings.

In the fall, when frost kills the plants, carefully dig up the tubers and cut off the stems 6″ above them. Let them dry in the sun for a day. Then wrap in newspaper or place in a box filled with peatmoss. Store in a dry cool (40°–50°) place. If tubers shrivel badly, sprinkle the peatmoss lightly with water occasionally.

DANDELION All-too-common weed which may be grown for greens.

Sow seeds in late summer in average soil in a sunny spot. Harvest the greens the next spring. The flavor is improved if you raise the leaves and tie them together loosely so that they are bleached. The danger in planting dandelions, of course, is that they may take over your place.

DAPHNE Deciduous and evergreen shrubs to 5′ with handsome foliage and fragrant flowers in various colors in spring or late winter. Most species grow best in warm climates but *D. cneorum* and *D. mezereum* are hardy fairly far north.

Propagate by layering. Young potted plants are most easily transplanted. Plant in spring or fall in average, somewhat sandy soil which must be well drained. *D. cneorum* is grown in the sun; others do best in partial shade. Avoid overwatering, but don't neglect plants in dry spells. Fertilize in the spring. Pinch out stem ends occasionally to develop bushier plants; otherwise, prune little, if any. In cold climates protect with straw in winter.

DAPHNIPHYLLUM MACROPODUM Broadleaf evergreen shrub to 25′ with handsome foliage. Grows in warm climates.

Plant in spring or fall in a sheltered spot in sun or partial shade. Give a rich, moist, lightly limed soil. Water in dry spells. Fertilize every 2–4 years.

DATE PALM *Phoenix.* Palms to 60′ with large crowns of feather-like fronds and edible fruits. Grows in warmest climates. Dwarf date palm (*P. roebelini*) is a good house plant.

For how to grow, see palm. However, note that while date palms are easily grown as ornamental trees, they bear fruit only in a few dry, very hot regions and then only (1) if a male tree is planted along with the female trees, (2) the weather is right and (3) considerable attention is given to the orchard.

DATURA Angel's trumpet, trumpet flower. Annual to 5′ with cone-shaped flowers in white, yellow or purple. Grows best in warm climates. Another angel's trumpet, *D. suaveolens*, is a tree with white flowers. It grows in subtropics.

Annual. Sow seeds indoors 8–10 weeks before last frost. Move outdoors into sun in rich, humusy soil. Fertilize when seedlings are well established and twice more during the summer. Water regularly.

Tree. Grow from seed. Plant in sun in rich, humusy soil. Water in dry weather. Fertilize in late winter every 1–2 years.

DAVALLIA Different species called rabbit's foot fern, squirrel's foot fern or ball fern. Delicate, feathery ferns with fronds to 18″ which sometimes droop, and creeping rhizomes that resemble animals' feet. A house plant.

For how to grow, see fern. *Davallias* are ideal for growing in hanging baskets. They are propagated in spring by dividing the rhizomes and fastening them on top of moist sand, covering lightly with sand or moss and keeping in a warm, dim place. Do not water too heavily while fern is resting in the winter.

DAWN REDWOOD *Metasequoia glyptostroboides.* Deciduous conifer to 100′ with upturning branches in pairs and bright-green foliage. Grows in warm and mild climate. See also *sequoia.*

Plant in spring or fall in light shade. Soil of average quality should contain considerable humus. Water regularly in all dry weather. Fertilize every 3–4 years.

DAYLILY *Hemerocallis.* Adaptable perennial to 5′ with lily-like flowers of yellow, orange, pink and red. Grows almost everywhere.

Divide roots in early spring or fall. To do this, first cut roots and tops back rather severely; then cut roots into sections with a knife. Grow in a sunny spot in average, well-drained soil that is deeply dug. Fertilize in early spring and again about 6–8 weeks later. Water regularly. Divide clumps every 4–5 years.

DELICATE LILY *Chlidanthus fragrans.* Bulbous plant to 10″ with yellow flowers in late summer. Grows almost everywhere.

Plant 3″ deep and 5″ apart in early spring in sun. The soil should be well drained, of average quality but with some extra humus. Fertilize once or twice. Water regularly. Dig up in the fall and store, when dried off, in peatmoss in a cool place.

DELPHINIUM Perennial larkspur. Perennial to 6′ with magnificent spikes of blue, purple, white or pink flowers in summer. Grows best in cool, moist climates.

Divide large clumps every third spring. Or sow seeds as soon as possible after they have been harvested in the summer and move plants to their permanent location in fall or early spring. The crowns should be set just above the soil surface. Grow in a sunny, airy spot. The soil must be a well-drained somewhat alkaline, rich loam with considerable humus. Apply a half cupful of general-purpose fertilizer in the spring. Water regularly. Stake the plants, especially the tallest varieties.

As soon as the first flowers fade, cut off the flower spike below the bottom flowers. Then when new growths are about 9″ tall, cut off the flower stalk at the ground and apply a little fertilizer to encourage a second flowering in the fall.

In winter, mulch plants with salt hay after the soil has frozen. To control insects and diseases, spray plants during spring and summer at weekly intervals with a combination insecticide-fungicide.

DENDROBIUM NOBILE Deciduous, epiphytic orchids to 30″ with groups of spring or early summer flowers in whites and purples. A house plant.

For how to grow, see orchid. During the plant's rest period, after flowering, water only enough to prevent pseudobulbs from shriveling, and give plenty of air (but not cold drafts).

DESERT WILLOW *Chilopsis linearis.* Shrub to 20′ with long, narrow leaves and lilac flowers. Grows in warm climates, especially in arid areas.

Plant in spring or fall in sun in average, well-drained soil. Once established, the plant should take care of itself.

DEUTZIA Deciduous shrubs to 8′ with many stems rising from the roots, and flowers, usually white but sometimes pink, in spring. Grows in all but coldest climates.

Propagate by green stem cuttings in spring, or by layering. Plant in spring or fall in sun or partial shade in average, well-drained soil. Fertilize every 1–2 years. Water in dry spells. Cut tired stems at the ground after plant has flowered.

DEWBERRY A form of blackberry but more trailing. Grows in warm climates.

Grow like blackberry (which see). However, canes should be trained to horizontal wires.

DIANTHUS Pinks, carnation, sweet William. Vast family of flowering plants—perennials, biennials and annuals to 3′. Except for greenhouse carnations, these grow almost everywhere, but prefer temperate climates.

Perennial. Divide plants in spring and plant in sandy, well-drained soil improved with a little humus and some fertilizer. If the soil is acid, it must be limed. Grow in full sun in an airy location. Water regularly but don't drown the plants. Spray weekly with zineb to control leafspot. In cold climates, cover with salt hay or evergreen boughs in winter. Divide crowded clumps every three years.

Biennial (sweet William). Sow seeds in middle to late spring and transplant to their permanent location in late summer or fall. Treat as above.

Annual. Sow seeds indoors 8–10 weeks before last frost. Transplant into a sunny spot in average, well-drained, limed soil. Fertilize when young plants are established. Keep watered. Pinch stems to promote bushy growth. Pick off dead flowers.

DICHONDRA CAROLINENSIS Perennial to 3″ with small, lily-pad-like leaves. Used as a grass substitute. Grows in warmest climates.

Sow seeds or plant sprigs in spring in sun or light shade. Grow in average soil prepared as for a lawn. Water in dry weather. Fertilize in spring. Mow occasionally or not at all, as you wish. Protect against fungus as you would grass (which see). Apply nemagon according to directions if attacked by nematodes. Weeds must be pulled by hand, since 2,4–D kills *dichondra.*

DICHORISANDRA Tropical perennial plants to 4′ with handsome leaves and blue flowers. A house plant.

Divide roots in winter and pot in equal parts of loam, humus and sand. Grow in a north window. Water regularly while plant is growing in spring and summer. Fertilize in early spring and at 6–8 week intervals thereafter until fall. When blooming ends, gradually withhold water and let plant die down.

DILL *Anethum graveolens.* Annual herb to 3′ with finely cut leaves and yellow flowers. Leaves and seeds are used for flavoring. Grows almost everywhere.

Sow seeds outdoors in spring after danger of frost is past. Dill needs sun, an average soil. Fertilize when plants are thinned to 18″. Water in dry weather.

DIMORPHOTHECA Cape marigold, cape daisy, African daisy. Annual to 1′ with daisy-like flowers in white, yellow, orange, pink. Grows almost everywhere.

Sow seeds indoors eight weeks before last frost. Move into a sunny location in average soil. Fertilize when young plants are established. Water in dry weather.

DIPELTA FLORIBUNDA Deciduous shrub to 15′, much like beauty bush, with profuse clusters of rose-and-orange, fragrant flowers in spring. Grows in mild and warm climates.

Plant in spring or fall in sun in average soil. Water in dry weather. Fertilize every 2–3 years. Prune after flowering.

DIZYGOTHECA Erroneously called *aralia.* Tropical shrubs kept to about 1′ indoors (but more out), with handsome foliage. A house plant.

Propagate by stem cuttings. Pot in equal parts of loam, humus and sand and keep moist. Grow in a warm north window. Spray foliage with water frequently and keep pot standing on wet pebbles. Fertilize every two months. Move outdoors in shade in summer. Spray with malathion to control scale.

DODECATHEON Shooting star. Perennials with leaves arranged in a low rosette, and flowers of various colors in a cluster on stems to 20″. Grows almost everywhere except in the south and east.

Divide plants in early spring or late fall and plant in sun or partial shade in average, well-drained soil containing a reasonable amount of humus. Water in dry weather, especially if plant is in full sun. Fertilize in spring.

DOGWOOD *Cornus*. Genus of deciduous trees, to 70′ but usually smaller, with profuse white or pink bracts in spring. There are also several unimportant shrubs. Different species are suited to different parts of the country, but none grows in coldest or warmest climates.

Plant in spring or fall in sun or partial shade. Grow in light, well-drained soil of average quality and preferably a bit acid. Fertilize every 2–3 years for best flower display. Water in dry weather. Prune lightly in winter to control shape. Remove suckers. Flowering dogwood (*C. florida*) is often attacked by a borer which is difficult to control; but it can be discouraged from attacking susceptible young trees by keeping the trunks wrapped in burlap for several years.

Flowering dogwood is easily propagated by planting the berries in the fall.

DOMBEYA One species called pink ball. Evergreen shrubs and trees to 30′ with beautiful, mainly pink flowers in winter. Grows in warmest climates.

Plant in spring in sun or partial shade in average soil. Water in dry weather. Fertilize annually. Protect from frost as much as possible, although if a plant is killed back, it should make rapid, sizable growth in the spring.

DOUGLAS FIR *Pseudotsuga taxifolia*. Conifer to 250′ with flattened needles and pendulous cones which have little tongues between the scales. Grows in the northwest and Rockies. The mountain variety can also be grown in the northeast.

Grow like fir (which see). Douglas fir grows in almost any soil that is well drained. It does well in partial shade as well as sun. Fertilize every year while young plant is getting started.

DOVE TREE *Davidia involucrata*. Deciduous tree to 50′ with large, hanging, creamy-white bracts in late spring. Grows in warm climates.

Plant in early spring or fall in sun in average soil to which a fair amount of humus has been added. Water in dry weather. Fertilize every 2–3 years.

DRABA Whitlow grass. Tufted perennial to 1′ with spring flowers in yellow, orange, white, or pink. Grows in rock gardens in mild climates.

Divide plants in spring or fall and plant in sun in average, well-drained soil. Water in dry weather. Fertilize lightly in early spring.

DRACAENA Trees or shrubs to 20′ but usually less, with handsome, rather narrow, striped leaves. Mainly a house plant but also grows outdoors in warmest climates.

Pot in general-purpose potting soil or the equivalent outdoors. Keep in an east or west window or in partial shade outdoors. Keep soil moist and fertilize monthly while plant is making growth. Spray foliage with water weekly to provide necessary humidity and discourage red spiders. Air-layer plants that become leggy, or take stem cuttings and throw the old plant away. To propagate by cuttings, chop a stem into 2″–3″ pieces, lay flat on damp sand and bury half way.

DRYAS Mountain avens. Spreading evergreen plants only a few inches tall with solitary yellow or white flowers. Grows in cold climates.

Plant in spring in a sunny position in average soil that must be well drained. Water in dry spells. Shade with evergreen boughs in winter.

DRYOPTERIS Different species called shield fern, wood fern, New York fern, Goldie's fern, oak fern, beech fern, marsh fern. Ferns to 3′, a few evergreen, of graceful but varying character. Grows in temperate climates but principally in the east.

For how to grow, see fern.

DUMB CANE *Dieffenbachia*, mother-in-law plant. Plant to 5′ with large, oval, upright leaves that are striped or mottled. Speech loss results if stem is chewed. Mainly a house plant, but also grows outdoors in warmest climates.

To propagate cut stem into 2″–3″ lengths, lay flat in moist sand, partly cover and keep warm until roots develop. Pot in general-purpose potting soil. Don't let soil dry out completely at any time. Spray leaves often with water, especially if house is very dry. Fertilize monthly. Needs only 2–3 hours of sun a day, even grows well in a north window.

DUMPLING CACTUS *Lophophora williamsi*, peyote. Spineless, round desert cactus 3″ tall, with small white or pink flowers. Grows indoors; also outdoors in hot, dry climates.

For how to grow, see cactus, desert. If potted, use a deep standard pot to accommodate the taproot.

DUSTY MILLER This name is applied to *Artemisia stelleriana*, *Centaurea cineraria* and *Lychnis coronaria*, but most often to *Centaurea gymnocarpa* and *Senecio cineraria*. Both of these 1′, white-foliaged perennials are grown in the north as annuals.

In the north, sow seeds indoors about ten weeks before last frost and move outdoors when weather is warm. In the south, sow seeds outdoors in the spring. In either case, grow in a sunny spot in average soil. Water regularly. Fertilize in spring.

DUTCHMAN'S BREECHES *Dicentra cucullaria*, white eardrops. Perennial wildflower 10″ tall with fern-like leaves and white flowers in early spring. Grows in the east.

Divide tuberous roots when foliage dies down after flowering, and plant 2″ deep in slightly acid, humusy soil that holds some moisture. Grow in shade and protect from wind. Give a little fertilizer every couple of years if you wish.

DUTCHMAN'S PIPE *Aristolochia durior*. Dense vine to 30′ with large leaves and hook-shaped, purplish-brown flowers. Grows almost everywhere.

Propagate by layering young growth. Plant in sun or shade in almost any location in average soil. Provide a strong support that the stems can twine around (vine becomes very heavy). Cut back hard in the spring and keep thinned out during summer. Spray with DDT to control caterpillars.

DYCKIA Terrestrial bromeliads to 3′ with sharp, spiny-edged leaves and tall orange flower spikes. Grows outdoors in subtropics; also indoors.

Propagate by offsets planted shallowly in one part sand and one part osmunda fiber. Water when soil feels dry, though these plants withstand some drought quite well. Fertilize with dilute liquid plant food every six weeks when making growth. Grow in the sun. Protect from frost. Spray with malathion to control scale, but rinse off next day with water.

EASTER CACTUS *Schlumbergera gaertneri.* Spineless jungle cactus with hanging branches about 2′ long and red flowers in spring. A house plant.

Grow like Christmas cactus (which see). The schedule, however, should be moved back two months; that is, keep dry in December, start watering and fertilizing in January.

EASTER LILY CACTUS *Echinopsis.* Cylindrical or round, ribbed, spiny desert cacti to about 2′ with large pink or white, lily-like flowers. Grows indoors; also outdoors in hot climates.

For how to grow, see cactus, desert. Outdoors, the plant will grow in sun or light shade. Indoors, it prefers full sun.

ECHEVERIA Different species known as hen-and-chickens, painted lady, Mexican firecracker, plush plant, chenille plant. Succulents to 2′, of different shapes but all with beautifully colored or textured leaves and some with handsome flowers. Grows indoors; also outdoors in hot, dry climates.

For how to grow, see cactus, desert. Mix a little extra humus into the soil. Water well when plant is making growth. Provide good ventilation for house plants and move them outdoors in summer. Smooth-leaved species should be exposed to full sun, but hairy-leaved species prefer partial shade or filtered sun. If roots develop swellings, prune them hard and dry off for several days before repotting in fresh soil.

ECHINOPS Globe thistle. Widely distributed perennials to 7′ with spiny foliage and globular, thistle-like flowers in various colors.

Divide plants in spring and plant in full sun in average soil that is deeply dug. Water sparingly. Fertilize lightly in the spring. Divide clumps every three years.

ECHIUM FASTUOSUM Pride of Madeira. Shrubby perennial to 4′ with greyish-green leaves and bluish-purple flowers in late spring. Grows in California's warm areas.

Propagate by stem cuttings in spring. Plant in spring or fall in sun in any well-drained soil. Water occasionally in dry weather. Fertilize lightly about every third spring.

EDELWEISS *Leontopodium alpinum.* Perennial to about 6″ covered with a whitish wool and with small yellow flowers. Grows in cold and mild climates.

Sow seeds indoors in winter and shift outdoors where plants are to grow when weather is warm. Flowering starts the next year. Grow in sun in average, well-drained soil. Water in dry weather. Fertilize the first year when plants are established outdoors; thereafter in the spring.

EGGPLANT *Solanum melongena esculentum.* Plant to 30″ with large, handsome purple fruits. Needs heat and grows best in warm and mild climates.

Eggplants require a better-than-average soil which is light, sandy, humusy, well drained. Start plants from seeds sown indoors about eight weeks before last frost, or buy plants. Space in the garden 2′–3′ apart each way. Grow in the sun. After the little plants are established, feed them with general-purpose fertilizer and make a second application a month later. Keep weeded. Water well in dry weather. To control black flea beetles that chew on leaves, dust with rotenone. Do not plant eggplants in the same place year after year.

EGYPTIAN PAPER PLANT *Cyperus papyrus.* A perennial rush with round, grassy heads like mops on stalks up to 8′. Grows in warm climates.

Divide roots in spring and plant in sun in average soil under 1″–3″ of water. Prune to keep from getting too tall.

ELAEAGNUS Different species called Russian olive, oleaster or Trebizond date; silverberry or wolfberry; lingaro. Shrubs or trees to 20′ with ornamental silvery foliage and yellow or red fruits. Different species grow in different areas, including coldest and warmest.

Plant in early spring or fall in a sunny spot. Grow in average, well-drained soil. Water in long dry spells. Fertilize every 2–3 years. Prune in early spring to control shape and size. Plants are easily trained or espaliered on a wall.

ELDER *Sambucus,* elderberry. Deciduous shrubs to 25′ with white flowers and berries often used for making wine or jelly. Grows almost everywhere.

Propagate by stem cuttings or suckers. Plant in partial shade in average, moist soil. Fertilize in spring. Prune in winter. Eliminate suckers.

ELEPHANT EAR *Colocasia esculenta,* taro. Tuberous plants to 9′ with excellent, edible foliage (leaves of some species are mammoth). Grows almost everywhere but is hardy only in warm climates. A house plant also.

Outdoors. Start tubers indoors in rich, humusy soil in early spring. Transplant outdoors in partial shade when danger of frost is past. Plant in rich, wet soil with the root crown level with the surface. Fertilize when plants are established and once or twice thereafter. In cold climates, store tubers indoors in a cool, dark place; in warm climates, leave them in the garden.

Indoors. Plant in equal parts of loam, humus and sand in a tub. Keep soil moist at all times. Fertilize monthly. Grow in an east or west window.

ELM *Ulmus.* Deciduous trees to 120′ with beautiful shape, oval leaves that turn yellow in fall. Grows almost everywhere.

The American elm, by far the best and handsomest of the genus, is fast being wiped out by the Dutch elm disease. Several other species are also affected by the disease. This leaves you with a choice of not planting elms or putting in one of the less good types, such as the Chinese elm.

Plant in spring or fall in sun in average, well-drained soil. Water in seriously dry weather. Fertilize every 3–4 years. (Fertilize yearly if you happen to have inherited an American elm on your property.) Clean out dead and broken limbs as they occur.

ELSHOLTZIA STAUNTONI Deciduous shrub to 4′ with spikes of lavender-pink flowers in the fall. Grows in mild and warm climates.

Plant in spring or fall in average soil in a sunny spot. Water in dry spells. Fertilize every 1–2 years in the spring.

EMILIA Tassel flower, Flora's paintbrush. Annual to 2′ with red or yellow flower clusters. Grows almost everywhere.

Sow seeds outdoors after danger of frost is past. Grow in sun in average soil. Fertilize when plants are thinned to 9″. Water in dry weather.

ENDIVE *Cichorium endivia*, escarole. A curled and fringed leafy vegetable to about 9″. Resembles lettuce. Grows almost everywhere. See also chicory.

Grow like lettuce (which see). To blanch endive, make a tarpaper roof over the plants when they mature. However, endive tastes just as good if you don't do this.

ENGLISH DAISY *Bellis perennis.* Six-inch plants with white, pink or red flowers in spring and early summer. A perennial but often treated as a biennial. Grows almost everywhere.

Sow seeds outdoors in a seedbed about August 1 and move plants into their permanent location in fall or spring. Mulch in winter to protect against very cold weather. Grow in partial shade in rich, well-drained soil. Fertilize in spring. Keep watered and cultivated. After flowering, plants can be propagated by division.

ENGLISH IVY *Hedera helix.* See ivy.

ENKIANTHUS Deciduous shrubs to 15′ with good foliage that turns brilliant colors in fall and white, red or yellow flowers in spring. Grows in temperate climates.

Propagate by green stem cuttings in summer. Plant in spring or fall in the sun. The soil should be well-drained, acid, and of average quality but with extra peatmoss mixed in. Water in dry spells. Fertilize every 1–2 years. Don't move plants once they are established.

EPIDENDRUM Very large group of epiphytic orchids to 4′ with large sprays of small flowers in various colors. A house plant. Also occasionally grown outdoors in southernmost Florida.

For how to grow, see orchid.

EPIMEDIUM Barrenwort. Perennials to 1′ with fine foliage and dainty spring flowers in yellow, red, white or blue. Grows in temperate climates.

Divide plants in spring or fall

and plant in light, humusy loam in partial shade. Keep watered and cultivated. Fertilize in spring. Old leaves help to protect plants in winter but should be cut off in spring to encourage new growth.

EPISCIA Trailing evergreen plants to about 15″ with white, red, orange or lilac flowers. Grows indoors.

Propagate by layering the runners or by removing them from the plant and rooting in moist sand and peatmoss. Pot plants in equal parts of loam, humus and sand. Keep moist. Fertilize every three weeks. Grow in a warm (about 65°) south window in a well-ventilated, but not drafty, room. Spray foliage with water daily. Move outdoors into partial shade in summer.

ERINUS Crevice plant. Perennial 4″ tall and grown in the rock garden in temperate climates for its small, purple, rose or white spring flowers.

Divide in spring or fall and plant in average soil that is very well-drained. Plant needs sun but should be shaded at midday in warm weather. Water sparingly in dry weather. Feed a little in early spring.

ERYTHEA Different species called blue hesper or grey goddess palm, San Jose paper palm, Guadalupe palm. Trees to 40′ with stout, rough trunks, fan-shaped leaves and pendent flower clusters. Subtropical.

For how to grow, see palm. The trees are rather slow growing.

ERYTHRINA Coral tree. Large genus of trees to 45′ with clusters of handsome red flowers. Grows in subtropics.

Plant in spring or fall in sun in average soil. Water in dry weather. Fertilize every other winter. Prune after flowering.

ERYTHRONIUM Dogtooth violet, trout lily, adder's tongue, fawn lily, Adam and Eve. Spring bulbs to 1′ with miniature flowers in white, yellow, rose and purple. Different species grow in most temperate regions.

Plant bulbs as soon as you receive them in a shady, moist spot where there is average, humusy soil. Set bulbs 3″ deep and about 3″ apart.

ESCALLONIA Handsome evergreen shrubs to 25′ with glossy leaves and clusters of white, pink or red flowers in late summer and fall. Grows in warmest parts of California.

Propagate by stem cuttings in the fall. Plant in spring in average soil. Grow in sun or light shade. Keep well watered while making growth, especially if plants are in the sun. Fertilize in spring. After flowering, remove flower heads and cut back old stems. Stems on large species can be reduced as much as one-third every year. Prune lower-growing hybrids only as necessary to control shape and growth.

EUCALYPTUS Huge genus of trees from 10′ to 200′ tall, of varying habit but generally graceful. Some with flowers. Grows best in warm parts of California.

Plant in spring in average soil in a sunny location. Take care not to damage roots. Many species can get by with little moisture in summer, but others require occasional deep watering. Ask plant dealer about the species you buy. Fertilize annually in the spring for the first several years, thereafter only every

3–5 years. Some of the smaller species, such as *E. pulverulenta* and *E. crucis*, should be lightly pruned in winter to encourage attractive new growth.

EUGENIA Different species called Surinam cherry, rose apple, Malay apple, jambolan, grumichama. Subtropical evergreen shrubs and trees to 40′ with small, edible but unexciting fruits. Surinam cherry is also a good house plant.

Outdoors. Plant in spring or fall in sun in average soil. Water copiously when fruit is developing. Fertilize every year. Smaller trees and shrubs respond well to pruning after they have fruited; consequently, their shape and size are easily controlled. Surinam cherry can be clipped like a hedge plant. Don't expect young plants to grow fast.

Indoors. Pot in general-purpose potting soil. Grow in a south window. Keep soil evenly moist and fertilize every two months. Move outdoors in summer.

EULALIA *Miscanthus sinensis.* Three-foot, ornamental, perennial grass with feathery panicles. Grows almost everywhere.

Divide plants in spring and plant in a sunny location in average soil.

EUONYMUS FORTUNEI Wintercreeper. Evergreen vine to 30′ with handsome and occasional variegated leaves and sometimes with showy, orange berry clusters. Grows in mild and cold (but not coldest) climates.

Propagate by stem cuttings of new growth or by layering. Plant in average soil, preferably with some humus, in sun or partial shade. Water in dry weather. Fertilize in spring. To control scale, which is a serious pest, spray with malathion or an oil emulsion in very early spring and twice again at monthly intervals. The vine grows by means of root-like holdfasts which cling to rough surfaces. Cut out wood badly infested with scale at any time. Prune otherwise in early spring.

EUONYMUS SHRUBS Different species called winged spindle tree; burning bush, strawberry bush or skewerwood; wahoo. Deciduous and evergreen shrubs to 20′ of different character, some with cork-like bark, some with pretty fruits, some with brilliant fall foliage. Deciduous types grow well in cold climates; evergreen types are hardy only in mild or warm climates. *E. japonicus*, an evergreen, also grows indoors.

Outdoors. Deciduous species often self-seed freely. The evergreens are propagated by young stem cuttings. Plant in spring or fall in average, well-drained soil in sun or partial shade. Water in dry weather. Fertilize when plants seem to need a shot in the arm. Prune hard in winter.

Indoors. Plant in general-purpose potting soil. Grow in a very cool east or west window. Water when soil feels dry. Fertilize once or twice a year. Move outdoors into light shade in summer.

EUPATORIUM Different species called Joe-Pye-Weed, boneset or thoroughwort, mistflower. Perennial wildflower to 6′ with flat clusters of fuzzy, not very brightly colored flowers in late summer or fall. Grows almost everywhere.

Divide plants in spring and plant

in sun or partial shade in average soil which retains moisture.

EUPHORBIA Vast family of succulent plants. Those referred to here are succulents which much resemble cacti. They are of many different forms, but have spines, milky sap and intricate flowers. Different species known as soldier, devil's club, cow's horns, milk tree. Grows indoors; also in warmest climates. See also crown of thorns, poinsettia, snow on the mountain.

For how to grow, see cactus, desert. Water sparingly while plants are resting. Protect from frost. Avoid wounding plants, because they bleed and become disfigured. The sap may also harm you if it gets into a cut, your eyes or mouth.

EVERGLADE PALM *Paurotis wrighti.* Shrub-like, 15′ palm with clustered trunks and smallish, fan-shaped fronds. Grows in southern Florida.

For how to grow, see palm. Give a constant supply of moisture.

EVERGREEN GRAPE *Cissus capensis.* Vigorous evergreen vine to 50′ with reddish-black fruits. Grows in warmest climates.

Propagate by stem cuttings and plant in spring or fall in sandy, humusy soil in partial shade. Keep watered. Spray with water occasionally to get rid of mealybugs. If necessary, use malathion. Provide a sturdy support. Keep pinched back.

EVODIA Deciduous trees to 25′ with handsome, aromatic foliage and panicles of white or pink flowers in late spring. Grows in temperate climates.

Plant in early spring or fall in sun in average soil. Water in dry spells. Fertilize every 2–3 years in spring.

FALSE CYPRESS *Chamaecyparis.* Also identified as *retinospora.* Different species called Port Orford cedar, white cedar. Conifers to 150′ with flat leaves. Grows in all but dry climates.

Plant in spring or early fall in average, well-drained, slightly acid soil which retains moisture well. Grow in sun but do not expose to strong winds. Water in dry weather. Fertilize every 1–2 years while young. Prune as necessary in spring just before growth begins.

FALSE DRAGONHEAD *Physostegia virginiana,* obedient plant. Weedy, four-foot perennial with spikes of red, pink, or white flowers in late summer. Grows in east and south.

Divide plants in spring or fall and plant in average soil in sun or shade. Water regularly in dry weather. Fertilize in spring.

FALSE INDIGO *Amorpha.* One species known as lead plant. Deciduous shrubs to 15′ with feathery foliage and summer flowers in blue or purple. Grows in mild and warm climates.

Propagate by stem cuttings in summer, or by layering. Plant in spring or fall in sun in average soil. Water in drought. Fertilize every 2–3 years. Prune back hard in early spring if plants get ratty.

FAME FLOWER *Talinum patens variegatum.* Two-foot plant with white-marked green leaves and red flowers. A house plant.

Pot in general-purpose potting soil. Grow in an east or west window. Water when soil feels dry.

Fertilize lightly 2–3 times in spring. Remove dead flowers.

FATSHEDERA A hybrid that climbs to about 5′ with large leaves like English ivy leaves. A house plant; also grows outdoors in warmest climates.

Pot in general-purpose potting soil. Grow in cool (about 65°) north window. Keep soil moist. Fertilize monthly. Spray foliage with water occasionally. Tie to a small trellis or stake. Keep stem ends pinched if you want a bushy plant. Move outdoors into shade in summer.

FATSIA JAPONICA Often identified as *Aralia sieboldi*. Evergreen shrub to 15′ with large, lobed leaves and clusters of small, white flowers. A house plant; also grows outdoors in warm climates.

Propagate by stem cuttings in spring. Plant in general-purpose potting soil or the equivalent outdoors. Grow in a north window or outdoor shade. Keep watered. Fertilize occasionally in spring, summer and (indoors) fall. Prune in winter to shape plant.

FEATHER DUSTER PALM *Rhopalostylis sapida*. Erect palm to 25′ with stiff, feather-like fronds branching out like a feather duster. Grows in subtropics,

For how to grow, see palm. Give partial shade.

FELICIA One species called blue marguerite. Perennials to 18″ with blue or pink flowers with yellow centers. Grows in warmest climates.

Sow seeds in early spring outdoors. Transplant into sun or partial shade in average soil. Water regularly. Fertilize once or twice while plant is making growth. Keep flowers picked off. Cut back hard in late summer to stimulate fall bloom.

FENNEL *Foeniculum vulgare*. Four-foot perennial herb, grown as an annual, with licorice-flavored leaves. Grows almost everywhere.

Sow seeds outdoors after danger of frost is past. Grow in sun in average, well-drained soil. Fertilize when plants are thinned to 6″. Water in dry weather.

FERN Huge group of feathery, flowerless plants. There are species of hardy ferns to be found in almost every moist climate. Both hardy and tropical ferns also grow indoors.

Outdoors. Propagate by division of the roots in late fall or early spring. Be careful how you handle the plants, since they are easily broken. Plant in the spring in light shade (but note that some ferns like sun and some also tolerate pretty deep shade). The soil should be well drained and contain considerable humus. A mixture of equal parts of humus, loam and sand is, as a rule, ideal. Don't crowd plants too close together because they need some air circulation, and don't bury the crowns. Water in dry weather. Cut off dead fronds in spring, when growth starts. Fall-planted ferns should be covered lightly with leaves the first winter.

Indoors. Pot in equal amounts of loam, humus and sand. Be sure pot drainage is good. Grow in a north window at a temperature of no more than 70° (and in winter somewhat less). Keep soil moist but not soggy at all times. To provide needed humidity, spray foliage daily with a fine water spray and stand pots on wet pebbles. Most species should be repotted every March in a little fresh soil. Feed monthly with weak liquid fertilizer while in active growth in spring and summer. To control scale (do not confuse this insect pest with the spore cases, which are clustered together neatly or arranged in lines), dip fern in a solution of nicotine sulfate and soap.

FEVERFEW *Chrysanthemum parthenium.* Perennial to 3′ covered with small yellow or white flowers. Grows almost everywhere if treated in the north as an annual.

Sow seeds outdoors in spring in full sun in average, well-drained soil. Water regularly and deeply. Fertilize when plants are thinned to 1′ and once more during early summer. Mulch in winter but don't count on plants' survival in the north. They may, however, self-sow.

FIDDLE-LEAF FIG *Ficus lyrata.* Tree to 10′ with long, fiddle-shaped, glossy leaves. Mainly a house plant, but also grows outdoors in subtropics.

Grow like rubber plant (which see).

FIG *Ficus carica.* Deciduous tree or shrub to 30′ with deeply lobed leaves and sweet, seedy, purple, green or brown, edible fruits. Grows best in warm climates.

Plant in spring or late fall in a sunny location on a north slope in deep, rich, fairly heavy soil that holds moisture. Space trees 15′–20′ apart. Water in dry weather. Apply general-purpose fertilizer in spring at the rate of 1 lb. per year of age. Cultivate shallowly. Prune out dead wood and suckers in spring. In cold climates, where they may be killed back, the trees should be bent to the ground and covered with soil and straw. Smyrna fig trees will not bear fruit unless pollinated by the fig wasp, which must be established in the orchard.

FILBERT *Corylus maxima.* Deciduous tree to 30′ with a decorative structure and oblong, edible nuts. Grows best in Oregon. See also hazelnut.

Plant in spring or fall in sun in average, well-drained soil. Two different varieties spaced 25′ apart are needed for pollination. Water in dry weather. Fertilize in spring. Prune in late winter to remove wood that bore nuts the previous year.

FILIPENDULA Meadowsweet, dropwort, queen of the prairie, queen of the meadow. Sometimes called spirea. Perennials to 7′ with large terminal clusters of white or pink flowers. Grows in temperate climates.

Divide clumps in spring and plant in sun or partial shade. Provide a humusy soil. Water regularly. Fertilize in spring.

FINOCCHIO *Foeniculum vulgare dulce.* Florence fennel, sweet fennel. Three-foot plant producing at the base a large, anise-flavored bulb which is cooked as a vegetable. Grows almost everywhere.

Grow like fennel (which see).

FIR *Abies.* Magnificent conifers to 300′ with short, flat needles and erect cones. Grows best in cool, moist, mountainous regions.

Don't try to grow firs in industrial atmospheres, and avoid exposure to strong winds. Plant in spring or early fall (except in very cold climates, where spring planting is always better). Grow in the sun in average, well-drained soil. Water well while young plants are getting established; thereafter, only in droughts. Fertilize every year for the first few years. Prune in spring if necessary.

FIRECRACKER VINE *Manettia.* Evergreen vine to 6′ with yellow-tipped red flowers in summer. A house plant; also grows in subtropics.

Indoors. Propagate by tip cuttings in spring. Pot in general-purpose potting soil which should be kept moist. Fertilize monthly. Place in a south window, but in summer screen lightly from midday sun. Keep pots on wet pebbles. Pinch plants until they are well-branched. Provide strings or wires for stems to twine around.

Outdoors. Same as above but grow in partial shade. Prune in early spring.

FIREWEED *Epilobium angustifolium,* great willow herb, rose bay, buckweed. Perennial wildflower with 4′ willow stems and rosy-purple flower clusters. Grows in mild and cold climates.

Divide plants in spring and plant in full sun in average, well-drained soil. Then leave alone.

FIREWHEEL TREE *Stenocarpus sinuatus.* Evergreen tree to 25′ with oak-like foliage and red flowers in a wheel around the ends of the stems. Subtropical.

Plant in spring or fall in sun. Soil should be well drained, of average quality, somewhat acid. Water in dry spells. Fertilize in winter every 3–4 years.

FITTONIA Perennial house plants to 18″ with handsome, heart-shaped leaves that are veined red or white.

Propagate in spring by stem cuttings. Pot in equal parts of loam, coarse humus, and sand. Grow in a warm north window. Keep soil moist. Fertilize 3–4 times annually.

FLAME VINE *Pyrostegia ignea.* Evergreen vine to 35′ with huge clusters of flagrantly orange, tubular flowers in summer and winter. Subtropical.

Propagate by stem cuttings. Plant in rich, well-drained soil with plenty of humus. Keep moist at all times. Fertilize 2–3 times a year. Flowers best when in full sun. Provide a sturdy support for vine to cling to. Prune hard in the spring right after flowering.

FLAMINGO FLOWER *Anthurium,* tailflower. Perennial to 3′ with long leaves and strange, long-lasting, white, pink, red or orange flowers more or less the year round. A house plant.

Plant in osmunda fiber in pots containing a deep layer of coarse drainage material. Keep roots covered with fiber as they grow (they tend to push out of the fiber). Water regularly and keep pots standing on wet pebbles. Fertilize in spring and January. Grow in a warm, east window. Repot only about every two years and then in January.

FLAX *Linum.* Annual flax (*L. grandiflorum*) grows to 2′, has red flowers. The best of the perennials (*L. perenne*) grows to 2′ and has blue or white flowers. Grows almost everywhere.

Sow seeds of annual flax in early

spring where plants are to grow. Perennial seeds can be sown up to midsummer in a seedbed and transplanted in fall or early spring to their permanent location. Flax needs full sun, average soil. Fertilize in spring. Water in dry weather.

FLEABANE *Erigeron,* dainty daisy, seaside daisy, beach aster, Robin's plantain. Perennials to 30″ with daisy-like, summer flowers in various colors. Grows in temperate climates.

Divide in spring and plant in a sunny position. Grow in average, well-drained soil which is watered moderately and improved in spring with a little fertilizer. Keep dead flowers picked off. Divide plants every third year.

FLOWERING ALMOND *Prunus glandulosa, P. nana* and *P. triloba.* Shrubs to 10′ with showy pink or white spring flowers. Grows in mild and warm climates. See also almond.

Grow like flowering crab (which see).

FLOWERING APRICOT *Prunus mume.* Deciduous tree to 30′ with fragrant, pink, spring flowers and very sour, plum-like fruits. Grows best in warmest parts of California.

Grow like flowering crab (which see). After flowering, cut back stems on which flowers were borne to 6″. (Flowers come only on two-year wood.)

FLOWERING CHERRY *Prunus.* Trees and shrubs to 25′, making a beautiful spring display of pink or white flowers. Grows in mild and warm climates.

Grow like flowering crab (which see). However, the cherries are not quite so hardy as the crabs and should be given protection against wind. Spray with dormant oil in winter to control scale insects.

FLOWERING CRAB *Malus,* crabapple, crabtree. Very ornamental deciduous trees and shrubs to 25′ with spring flowers in white, pink or red and sometimes edible fruits. Grows in all but most extreme climates.

Plant in early spring or fall in a sunny location. Grow in average, humusy soil that is well drained. Fertilize at planting and again a year later. Water well during the first year. Thereafter, do not feed except when the tree needs a rare shot in the arm. Water in very dry spells. Prune out dead and broken wood in winter. Spray with dormant oil in winter to control scale. Eliminate cedar trees on the property if crabs are infected by cedar-apple rust, which shows up in the form of orange spots on the leaves in summer.

FLOWERING CURRANT *Ribes.* Evergreen or deciduous shrubs to 12′ with pink or red flowers in spring. Grows in warm climates.

Be sure that any flowering currant you plant is not an alternate host for the white pine blister rust. Propagate by layering and plant in spring or fall in sun in average soil. Water in dry weather. Fertilize in spring. Prune after flowering.

FLOWERING MAPLE *Abutilon,* Chinese bellflower. Evergreen shrub to 10′ with showy, drooping flowers in various colors. A house plant; also grows outdoors in warmest climates.

Indoors. Sow seeds indoors in late winter; shift to small pots and move outdoors in sun in summer; then in fall, pinch back hard and bring indoors. Now pot in general-purpose potting soil and grow in a south window at a temperature not higher than 60°. Keep soil moist and feed every month. Move outdoors in summer. Cut back large plants in September to promote new growth and encourage bloom. Return to house.

Outdoors. Plant in humusy, well-drained, average soil in partial shade. Water regularly and fertilize 3–4 times a year when plant is making its best growth. Prune in fall to stimulate growth and blooming. Stem cuttings can be taken at that time if you wish.

FLOWERING PEACH *Amygdalus persica.* Often listed as *prunus.* Deciduous trees to 25′ with beautiful white or pink spring flowers. Grows in all but extreme climates.

Plant in spring in full sun in average soil. Water in dry spells. Fertilize every other year in spring. Cut back all but 8″ of previous year's growth right after flowering. Spray with dormant oil in winter if scale becomes a problem.

FLOWERING PLUM *Prunus.* Deciduous trees to 25′ with pink spring flowers. Grows in all but coldest climates.

Grow like flowering crab (which see).

FLOWERING QUINCE *Chaenomeles.* Many-branched, deciduous shrubs to 10′ with pink, red or white flowers in spring. Grows in mild and warm climates.

Plant in spring or fall in sun or light shade. Grow in average, well-drained soil. Water in long dry spells. Fertilize every 1–2 years in the spring. Prune after flowering.

FOAMFLOWER *Tiarella.* Evergreen wildflower with large toothed leaves and 10″ stalks with white or pink flower spikes in April. Grows mainly in the east.

Propagate by division or by the rooted runners. Plant in spring or fall in shade in moist, humusy, sandy, soil. Cut off runners to encourage bloom.

FORGET-ME-NOT *Myosotis.* Plants to 1′ with a profusion of dainty blue, white or pink flowers. Both biennial and perennial species are widely distributed throughout the country. The biennial is often sold as an annual.

Biennial. Sow seeds in early spring where they are to grow. Give sun or partial shade and an average soil. Keep moist. Fertilize lightly when plants are thinned to 4″. Plants self-sow readily.

Perennial. Sow seeds in spring for bloom the next year, or divide plants in the fall. Handle as above.

FORSYTHIA Golden bells. Deciduous shrubs to 12′ with sprays of bright yellow flowers in early spring. Grows in temperate climates.

Propagate by burying tips of growing stems in the soil (tip layering). Plant in spring or fall in sun in average soil. Water in dry weather. Fertilize every 2–3 years. Prune annually after flowering. Don't bob the top. The proper procedure is to cut some of the oldest stems completely to the ground.

FOTHERGILLA Deciduous shrubs to 10′ with thimble-shaped, white flower clusters in spring, and beautiful autumn foliage. Grows in mild and warm climates.

Plant in early spring or fall in average, well-drained soil which holds moisture. The shrub prefers sun but does well in partial shade. Water in dry weather. Fertilize every 2–3 years.

FOUR O'CLOCK *Mirabilis*, marvel of Peru. Perennial to 30″, grown as an annual, with tubular, red, yellow or white flowers opening late in the afternoon. Grows almost everywhere.

Sow seeds in average soil in early spring. Grow in the sun. Fertilize when plants are thinned to 18″ and once again about 4–6 weeks later. Water regularly. In the fall the tubers may be dug and stored like dahlia tubers (see dahlia).

FOXGLOVE *Digitalis*. Perennial to 4′ with massive spikes of tubular purple, yellow or white flowers. The outstanding species is the biennial, *D. purpurea*. Grows in all but most extreme climates, but prefers cool, moist climates.

Sow seeds in late spring or early summer and transplant to the garden in the fall. Mulch lightly in cold climates in winter. The soil should be well-drained and humusy, of average quality. Fertilize in early spring and once more 4–6 weeks later. Keep watered but don't allow soil to become soggy. Foxglove grows in sun or light shade. Start new seeds every year.

FOXTAIL LILY *Eremurus*, desert candle. Perennials to 12′ with spires of yellow, orange, pink or white, bell-shaped flowers. Grows in mild and warm climates.

Plant the brittle, fleshy roots in the fall in average, well-drained soil. Mulch with straw over winter. Fertilize in early spring and monthly thereafter during the growing season. Water regularly until flowering stops. Plants need full sun. Don't disturb those that are established.

FRANCOA RAMOSA Maiden's wreath. Perennial to 3′ with pink or white flower spikes in summer and fall. Grows mainly in California.

Divide in spring or fall and plant in average soil in sun or partial shade. Keep well watered while making growth. Fertilize in spring and again 4–6 weeks later.

FRANGIPANI *Plumeria*. Deciduous trees to 35′ shaped something like a candelabra and with very fragrant flowers in red, pink, yellow or white. Subtropical.

Propagate in summer by stem cuttings which are allowed to cure in the air for a day and are then inserted in moist sand. Plant in fall in sun in average soil. Water except in winter. Fertilize every 1–3 years.

FRANKLINIA Sometimes identified as *gordonia*. Usually deciduous trees to 30′ with large white flowers in late summer and scarlet autumn foliage. Grows in warm climates.

Plant in spring or fall in average, well-drained, humusy soil which retains moisture. Grow in sun or partial shade. Fertilize every 2–3 years in the spring.

FREESIA Plants to 10″ growing from corms and producing very fragrant flowers in many colors in winter and spring. Mainly house plants but also grown outdoors in warmest climates.

Indoors. Plant corms several to a pot at two-week intervals from about August 15 to November 30. Plant in general-purpose potting soil 1″ deep and water thoroughly. Then set in a cool (about 45°–50°), sunny place (a coldframe or sunporch, for example). Water sparingly until growth appears, then whenever the soil feels dry. Stake plants to keep them from flopping. Don't bother to save the corms when flowering ends: indoor growing doesn't help them any.

Outdoors. Plant corms in the fall about 2″ deep and 4″ apart in average, well-drained soil in the sun. (In Florida, to control nematodes, fumigate soil with methyl bromide 2–3 weeks before planting.) Water in dry spells. Fertilize lightly before flowering starts. Keep foliage growing after flowering stops by watering for a few weeks and feeding once. Then, after plants die down, dig up corms and store in peatmoss in a cool, dry place until fall.

FREMONTIA Flannel bush. Evergreen shrubs to 15′ with hairy leaves and yellow flowers in spring and early summer. Grows mainly in California.

Propagate by greenwood stem cuttings in spring. Plant in spring or fall in sun in well-drained, average soil. Do not water. Fertilize lightly in spring. Prune after flowering to keep plants shapely.

FRINGED POLYGALA *Polygala paucifolia.* Perennial, evergreen wildflower with 6″ tall leaves that turn red in winter and interesting magenta flowers. Grows mainly in cool climates.

Propagate by stem cuttings started in June. Plant in partial shade in moist, acid, humusy soil.

FRINGE TREE *Chionanthus,* white fringe, old man's beard. Deciduous shrubs or trees to 25′ with glossy foliage and loose, white, fragrant flower clusters in spring. Grows in mild and warm climates.

Plant in early spring or fall in full sun. Grow in sandy, average soil that holds moisture. Protect from wind in the north. Water in bad dry spells. Fertilize every 2–3 years. Prune after flowering.

FRITILLARIA Guinea-hen flower, checkered lily. Bulbous plants to 4′ with hanging, bell-shaped flowers in various colors in spring. Grows in cool climates.

Bulbs of the small types should be planted 3″ deep in well-drained, average soil which will not dry out completely. Grow in light shade. Water in drought. The species known as crown imperial is planted 4″–5″ deep in somewhat richer soil. Fertilize every other spring. Don't disturb unless bloom becomes poor.

FUCHSIA Lovely shrubs to 20′ with pendulous flowers in shades and combinations of red, purple, blue and white. Grows outdoors in warm, moist climates. Also a house plant.

Outdoors. Propagate by softwood stem cuttings in spring. Plant in spring or fall in filtered sun in a protected spot. The soil should be light, rich, very well drained, slightly acid. To increase acidity mix a large amount of peatmoss into your garden soil and add considerable well-rotted manure if available. Keep soil moist at all times. Spray foliage with water frequently, especially on windy and hot days. Fertilize every 1–2 weeks with a small amount of general-purpose plant food (frequent small doses are better than occasional large doses).

Prune plants in spring before they start into growth by removing all weak branches completely and cutting back strong branches to two buds. Potted plants should be root-pruned at this time. Thereafter, throughout the spring, summer and fall, remove excess branches and cut back others to control and direct growth. Pinch stem ends to promote bushier specimens. Or you can grow the plant with a few stems which are trained up a wall or trellis.

Indoors. Repot old plants, after they have been pruned, in January; or pot up new young plants. Plant in well-drained pots in general-purpose potting soil containing extra humus. Grow in an east or west window at a temperature of no more than 60° (lower at night). Keep soil moist. Spray foliage with water frequently. Feed every two weeks with half-strength liquid plant food. Until March 30, pinch out stem ends to promote bushiness. Move outdoors in summer after danger of frost is past. Grow in filtered sun.

FURZE Ulex, gorse, whin. Much-branched, spiny shrubs to 3′, virtually leafless but with showy yellow flowers. Grows in warm climates.

Sow seeds where plants are to grow in the sun. The soil should be spare, sandy or gravelly; otherwise plants do not flower well. Grow furze near the sea or elsewhere in sandy country. It needs little attention.

GAILLARDIA Blanket flower, gay flower. Annual and perennial plants to 2′ with long-stemmed, single and double flowers in yellow, orange, red or white. Grows almost everywhere.

Annual. Sow seeds indoors 6–8 weeks before last frost or outdoors where plants are to grow when weather is warm. Grow in full sun, average soil. Water in dry weather. Fertilize when young plants are thinned to 1′ or established in garden.

Perennial. Propagate by division in the spring or sow seeds in a seedbed when weather is warm and transplant to their permanent location when plants are large enough to handle. Otherwise, grow like the annuals.

GALAX APHYLLA Beetleweed. Perennial, evergreen wildflower with round, leathery leaves and white flower spikes to 30″. Grows in the middle south.

Divide in early spring and plant in partial shade in somewhat acid, humusy soil that is well-drained and moist.

GARDENIA *Gardenia jasminoides*, cape jasmine. Handsome evergreen shrub to 6′ with waxy, white, very fragrant flowers in spring and summer. Grows in warm climates; also indoors.

Outdoors. Grow like *camellia* (which see). However, gardenias should be grown in a sheltered, sunny position. Fertilize monthly from March through September. Pick flowers while dew is still on them.

Indoors. Pot in a mixture of two parts loam, two parts peatmoss and one part sand to which some aluminum sulfate has been added. Keep soil evenly moist but not soggy. Fertilize monthly from March through September. Grow in a south window at a daytime temperature of 70°–72° and a nighttime temperature of 62°–65°. Spray foliage with water frequently, and stand pots on wet pebbles. Move outdoors in summer and plunge pots in filtered sun. Keep watered and sprayed.

GARLIC *Allium sativum.* Onion relative used for flavoring. Grows almost everywhere.

Plant the individual cloves ½″ deep and 4″–6″ apart in rows 1′ apart in average soil. The planting is made in the spring as soon as the soil can be worked. Grow in the sun. Keep free of weeds and water in dry weather. Dig up plants in the fall when the leaves die and hang in an airy, dry place.

GAS PLANT *Dictamnus,* dittany, fraxinella, burning bush. Perennial to 3′ with strong fragrance and upright white flower clusters. The plant gives off a gas which can be ignited. Grows almost everywhere.

Sow seeds in the fall outdoors. Germination occurs next spring, but don't transplant the seedlings for another year. Then place in the garden in a sunny location from which the plants should not be moved. Give plenty of space. The soil should be deeply dug, well-drained and rich. Fertilize in spring. Water sparingly.

GASTERIA Ox tongue. Succulents to 4′ with tongue-shaped leaves in rosettes and pendent flowers in red shades. Grows indoors; also in hot, dry climates.

For how to grow, see cactus, desert. Grow in light shade in summer.

GAULTHERIA *G. procumbens* known as partridge berry, wintergreen, checkerberry, teaberry, spiceberry or ground holly. A low, creeping evergreen with small, red berries, growing in cool and cold climates. *G. shallon,* called salal, is a spreading evergreen shrub to 10′, growing in warmer climates in the west.

Buy plants; transplanting from the wild is difficult. Plant in spring or fall in light shade in well-drained soil that contains considerable sand and peatmoss. Water in dry weather. If planted in sun, salal is used as a 1′ groundcover; in shade with moisture it grows tall.

GENISTA Broom. Deciduous and evergreen shrubs to 20′ with clusters of yellow or white flowers in spring. Grows in warm and mild climates. The *genista* sold by florists is actually *Cytisus canariensis* and is grown only in greenhouses.

Propagate by layering. Plant in spring or fall in a sunny spot in average, well-drained soil. The plants need little water except in drought. Fertilize lightly in early spring. Prune a little in summer and pick off dead flowers.

GENTIAN *Gentiana.* Large genus of perennials, biennials, and annuals to 4′ with flowers in many shades of blue but sometimes in other colors. Grows mainly in cool climates.

Gentians defy even the most expert gardeners. Better leave them alone.

GERANIUM *Pelargonium.* Ever-popular perennials to 6′ with handsome leaves often colored and scented, and with showy flowers in white, pink, red, orange or purple As a pot plant, geranium grows everywhere. Also grows outdoors the year round in frost-free climates, notably California.

Zonal geraniums. Don't try to carry geraniums over from year to year. For one thing, they do not bloom continuously; in fact, they take a pretty long rest every year. For another thing, the plants get large and have to be cut back fairly hard to force them to renew their vigor. Except in warmest climates, the best way to handle geraniums is to keep new plants coming along for cold-weather and warm-weather bloom. Assuming you have some geraniums growing outdoors in the summer, here's how you do this:

Take new stem cuttings from your outdoor plants in August. The cuttings should contain five nodes. Remove the leaves from the bottom three nodes and insert that part of the stem in a moist mixture of sand and peatmoss under polyethylene film. Keep outdoors in the shade until roots are well formed. Then pot the plants in a mixture of three parts loam, one part humus and one part sand with a little bone meal added if you wish. The pots must contain a good layer of coarse drainage material. Start the young plants in 2″ pots and repot in 3″, 4″, 5″ pots, etc. as the plants grow. Geraniums do best when they are somewhat rootbound, so don't repot them until the roots come out the drainage holes or the plant growth slows down.

Water geraniums thoroughly (until water flows out the bottom of the pot) but only when the soil has dried a little way below the surface. (In other words, the soil surface is dry; and about ½″ of the soil below that is dry.) Once plants are well rooted, fertilize lightly every month (except from November through February, when the frequency should be every six weeks). Any plant food will do but one that has a low nitrogen content is preferred.

While the potted plants are outdoors, keep them in the sun. Before frost, bring them indoors and set them in a south window. The room should be well ventilated, but don't expose the plants to cold drafts and don t grow them too close to cold window glass. Spray foliage monthly with water to keep it clean. To promote bushier plants and discourage legginess (which is caused mainly by lack of sun), pinch one or two of the side stems very occasionally. Don't, however, pinch too many branches, lest you delay blooming. Properly handled, plants which are started from cuttings in August should start blooming indoors in October.

For summer plants, take cuttings from your house plants in January or February. The technique of propagating and handling is the same as above, except that of course it is done indoors. Move potted plants outdoors as soon as danger of frost is past. Plants can be kept in pots all summer or planted in the garden in average, well-drained soil. Grow in full sun and continue watering and feeding as above. Take cuttings for the next winter's bloom in August.

As previously noted, the above procedure is the best way to get almost continuous geranium bloom. However, it is true that, if you hate to throw out beautiful large house plants when summer comes, you can cut them back somewhat, repot in fresh soil and move outdoors into sun. They will do pretty well, but eventually they will grow tired and stop blooming. Similarly, if you bring beautiful summer-blooming plants indoors in the fall, they will stop blooming—almost immediately—and they won't resume for months.

Common geranium problems are handled as follows: For insect pests, spray with malathion. For botrytis, which makes leaves go limp and develop greyish-brown spots, spray weekly with captan. If plant stems become watery and foliage wilts, the plant has bacterial stem rot—a very infectious, easily spread disease—and should be destroyed immediately.

Colored-leaved geraniums. Handle like zonal geraniums. Outdoors, these plants prefer hot, fairly dry climates.

Scented geraniums. Handle like zonal geraniums. The soil mixture should contain an additional half part of humus.

Ivy geraniums. Handle like zonal geraniums. Fertilize not more than twice, and then sparingly, from October to February; then fertilize every three weeks until plants can be moved outdoors.

Lady Washington geraniums. These are difficult indoor plants and better left alone. For outdoor culture, see below.

Tree geraniums. The care is the same as for zonal geraniums but you should fertilize more often. To develop a tree geranium, pot a tall, vigorous, rooted cutting in a 6″ pot. Set a 3′ stake close behind it and tie the plant to it loosely. Nip off all leaves except the four at the top of the stem. As the plant grows, remove all leaves except four at the top. Continue to tie the stem to the stake. When the plant is about 30″ tall, pinch the stem tip. Branches will now begin to develop at the top of the stem.

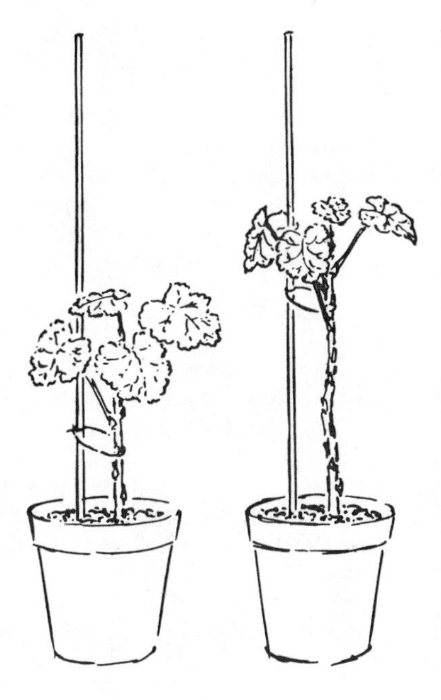

Geraniums outdoors the year round. Grow in pots or in beds. Soil and watering are the same as above. Plants grown in beds usually need less frequent applications of fertilizer than those in pots. Grow in sun, or in partial sun in hottest, driest climates (such as the interior valleys of California). A more or less constant supply of new plants can be developed by cuttings taken in August or March or whenever plants are making vigorous growth; or you can carry over old plants from year to year by cutting them back rather severely once a year. In the case of most geraniums, the best time for this hard pruning is the spring. In the case of Lady Washingtons, however, fall is preferred (do the job gradually from late August on). In addition, you should also pinch stem ends, remove faded flowers and cut out awkward stems of all geraniums more or less constantly throughout the growing season.

Protect plants from frost in winter by covering with newspapers, burlap or baskets or by moving potted plants under cover.

GERBERA Transvaal daisy, African daisy. Warm-climate perennial to 30″ with daisy-like flowers in red, orange and yellow.

Divide plants in late winter and plant in sun or partial shade in well-drained, deeply dug, humusy soil which should be limed if it is too acid. The crowns of the plants must be level with the soil surface. Keep watered during the active growing season. Avoid splashing leaves. Fertilize every 6–8 weeks from midwinter until growth slows in the fall. Apply a winter mulch in areas with frost.

GERMAN IVY *Senecio mikanioides.* Dense vine like English ivy, to 10′, with tiny yellow flowers in winter. A house plant. Also grows outdoors in warm climates.

Indoors. Propagate by division of the roots or by stem cuttings. Pot in general-purpose potting soil which is kept moist. Fertilize in the fall. Keep in a cool east or west window. Provide a light support for vine to climb on. Prune after flowering.

Outdoors. Plant in soil with extra humus in partial shade. Handle as above. Vine is killed to the ground by frost but quickly makes new growth.

GEUM Avens. Perennials to 2′ with bright red, orange or yellow flowers. Grows best in cool climates.

Divide in spring or sow seeds up to midsummer outdoors. Grow in sun or partial shade in well-drained, humusy soil which is enriched with fertilizer in early spring and again about June. Water regularly during the growing season. Mulch in winter.

GHERKIN A small cucumber for pickling and relishes. See cucumber.

GILIA Different species called bird's eyes, fairy stars, Queen Anne's thimble. Annuals to 2′ with flower clusters in many colors. Grows mainly in California.

Sow seeds outdoors in early spring (or in fall in areas with mild winters). Grow in sun in average soil. Fertilize when plants are thinned to 9″. Water in dry weather.

GINKGO Maidenhair tree. Handsome deciduous tree to 120′ with unusual fanshaped leaves. Grows in mild and warm climates.

Plant male trees only; the females bear foul-smelling fruits. Gingko is very tolerant of city conditions. Plant in early spring or fall in sun or partial shade in average, well-drained soil. Water deeply in drought. Fertilize young trees every year.

GLADIOLUS Plants growing from corms to 4′, with flower spikes in many colors. Grows anywhere.

Plant the corms in a sunny location in any good, well-drained soil. For a succession of bloom, make plantings at 2-week intervals from last frost through June. Set the corms 6″ deep instead of the 4″ usually recommended. This helps to prevent the tall plants from tipping over (you can also hill soil up around the stalks a few inches). Don't let the soil dry out completely. Apply general-purpose fertilizer with a low nitrogen content before, during, and after flowering. To keep down thrips, which may be a serious nuisance, spray plants with malathion three times at ten-day intervals after the plants are 8″ tall or apply a systemic poison.

Dig up the corms in the fall after the foliage has died. Do not pull off stems until they are easily separated from the corms by a light tug. Let the corms dry, then separate the old corm from the new one and harvest the little cormels for planting in a nursery bed the next spring. Store in uncovered trays or boxes in a dry, cool (40°–50°) place.

GLOBE AMARANTH Gomphrena, bachelor's button. An annual everlasting to 20″ with clover-like, red, pink, white or yellow flowers. Grows almost everywhere.

Sow seeds where the plants are to grow after frost danger is past. Grow in full sun in average soil. Water in dry weather. To dry, hang head down in a shady spot.

GLOBULARIA Globe daisy. Perennials to 1′ with blue or white, globe-shaped, spring flowers. Grows in mild and warm climates.

Divide in early spring and plant in partial shade in average, well-drained soil. Water in dry weather. Fertilize in spring.

GLORIOSA Glory lily. Tuberous-rooted vine to 10′ with lily-like flowers of red or yellow. Grows almost everywhere. Also a house plant.

Outdoors. After danger of frost is past, plant tubers 4″ deep (lay them flat) in average soil to which some humus has been added. Grow in full sun. Water well at the start, then sparingly until growth appears, then plentifully until flowering stops. Fertilize every fortnight after growth appears. Provide support for tendrils of plant to cling to. When flowering ends, stop feeding and gradually withhold water. In the fall, before soil freezes, lift tubers, cut off foliage and store like dahlia tubers (see dahlia).

In areas where ground doesn't freeze, tubers may be left outdoors the year round. When foliage dies, just tug it gently from the tuber.

Gloriosa tubers multiply by forming V's. To propagate, cut through the point of the V.

Indoors. Plant tubers in general-purpose potting soil in large pots in February or March. Set the tubers on a slant, with the eye at the top, and cover the eye with 2″ or more of soil. Place in a sunny window. Water well to start, then just enough to keep the soil from drying out. When growth appears, however, keep soil constantly moist and feed every two weeks with liquid plant food. Move outdoors in summer. After flowering stops, stop feeding and withhold water gradually.

When foliage dries, unpot tuber and store as above.

GLORY FLOWER *Eccremocarpus scaber.* Perennial vine to 12′ with large clusters of yellow, orange or red flowers in summer. Grows in warmest climates but can be grown in the north if treated as an annual.

If grown as an annual, sow seeds outdoors where plants are to grow when danger of frost is past; or sow seeds indoors about eight weeks before last frost. Grow in a well-drained soil enriched with humus, general-purpose fertilizer, and lime. Keep watered and fertilize plants several times during the summer. Grow in full sun or light shade. Protect from strong winds. Train on a light trellis. Stems intertwine badly if not controlled.

If glory flower is grown as a perennial, cut it back in late fall or early spring.

GLORY GRAPE *Vitis coignetiae,* glory vine. Fast-growing, deciduous grape vine to 65′ with big leaves that turn scarlet in the fall and inedible black fruit. Grows in all but coldest climates.

Grow like edible grapes. See grape. Give plants sturdy support. Prune in fall or spring to restrain and direct growth.

GLORY OF THE SNOW *Chionodoxa.* Spring bulbs only a few inches tall with clusters of blue flowers. Grows best in cool climates. Also indoors.

Plant in average, well-drained soil in the sun in the fall. Set bulbs 3″ deep and about 2″ apart. To grow indoors, see How to Force Bulbs for Indoor Bloom.

GLORY OF THE SUN *Leucocoryne ixioides.* Bulbous plant 1′ tall with clusters of blue flowers with white throats. Grows in southern California.

Plant in fall in sun in average, well-drained, sandy soil. Bulbs should be 3″ deep, about 5″ apart. Water while making growth. Fertilize in spring. Keep foliage growing for a while after flowering stops.

GLORY VINE *Clytostoma callistegioides,* violet trumpet vine, pointed trumpet. Evergreen vine to 30′ with lavender flowers. Subtropical.

Propagate by layering. Plant in spring or fall in sun in average soil. Water in dry weather. Fertilize in winter and spring. Prune after flowering in spring. Provide a trellis which tendrils can attach themselves to. This plant can also be used as a groundcover.

GLOXINIA *Sinningia speciosa.* Tuberous plant to 15″ with large, furry leaves and spectacular, bell-shaped flowers in white, pink, red or purple. A house plant.

Propagate by leaf cuttings in water or moist sand, or by cutting the tubers into two pieces each with a sprout and dusting the cut surfaces with sulfur. Pot in a sterilized mixture of equal parts of loam, humus and sand. Set the tuber so that the top just shows above the soil. Keep warm and in a bright but not sunny place. The best time to plant gloxinias is February, but you can do it at any time if tubers have had a couple of months' rest.

When growth starts, keep plant in an east or west window or a south window that is lightly screened at midday. Stand pots on wet pebbles. Water thoroughly when soil feels dry. Feed with general-purpose fertilizer every three weeks until flowering stops. If tuber puts up several stems, remove all but one of them.

To control cyclamen mites, which cause plants to be stunted and leaves to curl, set pots in a sodium selenate solution until soil is wet through. Spray with malathion to control thrips, which cause leaf tips to wither and become spotted underneath.

After flowering stops, gradually reduce water and stop fertilizing. When plants die down, remove the foliage and store the tuber, still in its pot, in a cool, dark place for 2–3 months. When restarting into growth, replace upper inch of soil in pot with new soil. Repotting is necessary only when tuber obviously becomes too large for pot.

GODETIA Satin flower, farewell to spring. Annuals to 2′ with showy, satiny flowers in white, pink, violet. Grows best in the Pacific states and other cool climates.

Sow seeds outdoors in spring when weather is reliable. Grow in partial shade in average soil. Keep moist. Fertilize when plants are thinned to 9″ and once more during the summer.

GOLD DUST PLANT *Aucuba japonica variegata.* Evergreen shrub to 10′ with leaves splattered with yellow, and bright-red fruits in fall *A. japonica,* with solid green leaves, is also called Japanese laurel. Grows in warm climates; also indoors.

Outdoors. Propagate by stem cuttings in spring or summer. Plant in spring or fall in shade in average, well-drained soil which should be kept moist. Fertilize every 2–3 years. To assure fruiting, plant both male and female specimens. Prune in winter.

Indoors. Pot in general-purpose potting soil and grow in an east or west window in a very cool (down to 45°) room. Keep watered. Fertilize about three times a year. Repot annually.

GOLDENCHAIN TREE *Laburnum.* Deciduous tree to 20′ with large, pendulous clusters of yellow flowers in spring. Flowers, foliage and fruits are poisonous. Grows in warm and mild climates.

Plant in early spring in light shade or sun. Grow in average soil improved with a little lime. Water in dry weather. Fertilize every 2–3 years in early spring.

GOLDEN CHINQUAPIN *Castanopsis chrysophylla minor.* Ornamental, evergreen shrubby tree to 15′ with leaves that are green on top, golden beneath. Grows in California and Oregon.

Plant in spring or fall in sun in average, well-drained soil. Water occasionally and deeply in dry weather. Fertilize every 3–4 years. Prune in late winter to control shape.

GOLDEN DEWDROP *Duranta repens,* sky flower, pigeon berry. Sprawling evergreen shrub to 15′ with lavender flowers and yellow berries. Grows in warmest climates.

Propagate by layering. Plant in spring or fall in a sheltered, sunny position in average soil. Water in dry weather. Fertilize every 1–2 years in late winter. Prune to control growth and thin out wood after fruiting.

GOLDEN LARCH *Pseudolarix amabilis* or *kaempferi.* Deciduous conifer to 100′ with rosettes of flat green needles which turn yellow and then drop in the fall. Grows in temperate climates.

Plant in early spring or fall in a sunny spot. Grow in average, well-drained, moisture-retentive soil. Fertilize every year or so while getting established. Water in dry weather and in fall before the ground freezes.

GOLDENRAIN TREE *Koelreuteria paniculata,* China tree, varnish tree, pride of India. Dome-shaped, deciduous tree to 30′ with large clusters of yellow summer flowers. Grows in mild and warm climates. A larger species, *K. formosana,* grows in subtropics.

Plant in early spring or fall in full sun in average, well-drained soil. Fertilize every 2–3 years. Withstands drought.

GOLDTHREAD *Coptis trifolia,* yellowroot, canker root. Also identified as *C. groenlandica.* Creeping, evergreen wilding, only a few inches tall, with small white flowers and thread-like, yellow roots. Grows in cold climates.

Divide in spring and plant in shade in very acid, moist, humusy soil.

GOOSEBERRY *Ribes grossularia.* Prickly deciduous shrubs to 5′ with green, yellow or red berries filled with seeds. Grows in cool, moist climates.

Note. Gooseberries should never be planted anywhere near white pines, because they are alternate hosts to the white pine blister rust. Some states prohibit planting of gooseberries.

Grow like currants (which see). Gooseberries, however, do best in light shade, because they dislike summer heat. Thin out some of the small side branches when pruning out the main stems as for currants.

GOURDS Plants belonging to various genera, trailing or climbing several feet and producing inedible fruits of various shapes, colors and sizes. Grows almost everywhere, but needs about three months of warmth.

The soil should be well drained, humusy and fertile. Full sun is required. Sow seeds 1″ deep in "hills" 3′ apart as soon as all danger of frost is past and the soil is warm. (In very cold climates, seeds can be started indoors in pots and the plants moved out when the ground is warm.) Thin to one plant per hill. Provide a trellis, about 3′ high, of chicken wire for the vines to climb on. Keep soil weeded and lightly cultivated. Do not let it dry out for long. Fertilize at the end of the first and second months. Dust plants frequently with rotenone to control cucumber beetles. If gourds become hidden by leaves, remove a few of the leaves to expose fruit to sun. Prune back vines if they grow too long. Do not remove the gourds from the vines until they are hard to the fingernail. However, pick before frost. Keep in a dry place at about 70° for three weeks, then move to a cooler place. When the fruit gives off a hard ring when tapped, it can be varnished, shellacked, painted or waxed.

GRAPE *Vitis.* Deciduous vines to 50′, with clusters of delicious fruits.

Grows almost everywhere except in desert regions.

Plant in early spring in a sunny location in deep, porous, well-drained, average soil. In the orchard, space vines 7′ apart in rows 7′ apart; around an arbor, space them 4′ apart. Set plants 2″ deeper than they formerly grew. Keep soil cultivated and weeded. Water in dry weather. During the first three years feed each plant in the early spring with nitrate of soda. Give a cup the first year; two cups the second; four cups the third. Thereafter, apply two cups of general-purpose fertilizer every year.

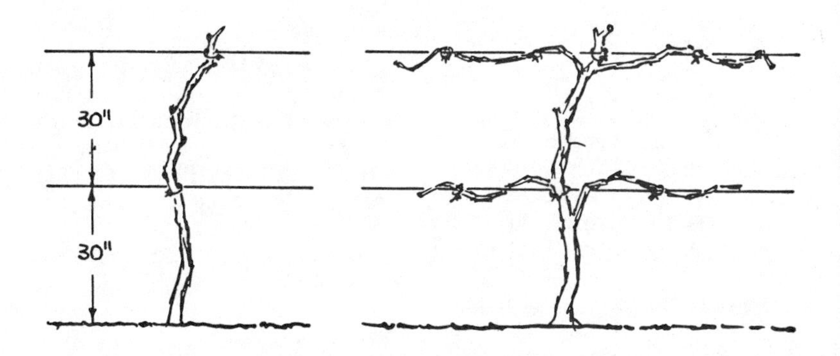

If growing grapes in an orchard, train them to two strong wires stretched tightly between stout posts, one wire at a height of 30″ above the ground, the other 30″ above that. If growing on a trellis, make sure the trellis is sturdy.

Grapes are pruned when dormant. In the orchard, cut back newly planted vines to one stem with 2–3 buds. Train the stems straight up and tie loosely to the wires. The next year remove all but one stem and cut this back to the top wire or even lower. This stem will form the trunk. The third year select two pencil-sized branches on either side of the trunk at the top wire and two other branches on either side of the trunk at the bottom wire. Cut these back to four buds and tie along the wires horizontally. Remove all other branches, but leave a single bud on the trunk close to each branch. These buds will develop into branches. The next year, cut out the old branches; trim back the year-old, pencil-size branches to 12 buds and tie to the wires on either side of the trunk. Again leave a single bud near each branch. From now on, repeat this system. The best grapes are borne on year-old, pencil-size growths.

If grapes are trained to a trellis, train plants to a single trunk, but instead of removing most of the branches, cut them back 25 per cent each year until the trellis is covered. Thereafter, thin out old and very small branches each year, and leave year-old, pencil-size branches. Cut these back to 12–15 buds.

The Muscadine types of grapes, such as scuppernong, are usually trained on large trellises. A single

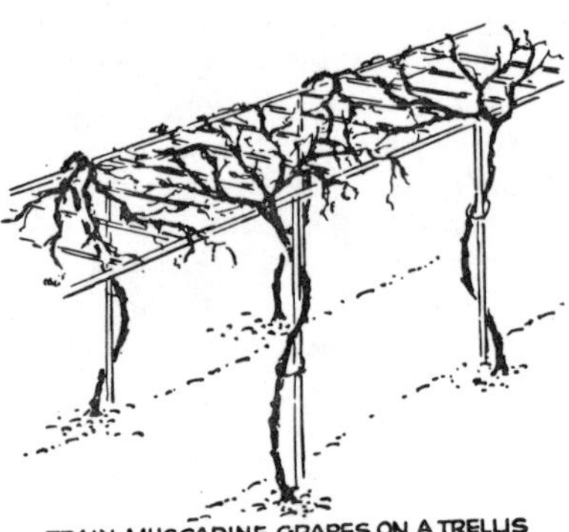
TRAIN MUSCADINE GRAPES ON A TRELLIS

stem, or trunk, is grown to the top of the trellis. Six to eight branches are then grown outwards from that point. Branches growing off these main branches are cut back to 2–3 buds each year.

To control insects and disease on grapes, spray as follows: Apply Bordeaux mixture in early spring when new growths are 1″ long, and again when shoots are 10″ long. Then apply general-purpose fruit spray as soon as tiny grapes start to form, three weeks later, and a month before grapes are ready for harvesting.

GRAPEFRUIT *Citrus paradisi.* Evergreen trees to 25′ with white flowers and large yellow fruits. Grows in warmest climates.

For how to grow, see orange. Space trees 35′ apart. Pruning of low-hanging branches is especially important since they may be weighted very low by the heavy fruit and thus give mealybugs and scale insects all too easy access to the trees from the ground.

GRAPE HYACINTH *Muscari.* Bulbs to 9″ with plump spikes of usually blue spring flowers. Grows almost everywhere. Also indoors.

Plant in fall in average, preferably gritty, soil in a sunny location. Plant 3″ deep and 3″ apart. To grow indoors, see How to Force Bulbs for Indoor Bloom.

GRAPE IVY *Cissus rhombifolia.* Fast-growing, glossy-leaved vine to 20′. Mainly grown indoors, but may also be grown outdoors in warm climates.

Indoors. Pot in general-purpose potting soil. Place in an east or west window. Keep soil moist. Spray foliage with water every week to control mealybugs. Pinch stem ends occasionally to keep vine in bounds. Fertilize 3–4 times a year.

Outdoors. Grow like evergreen grape (which see).

GRASS Whether your lawn is a pride and joy or the bane of your existence depends entirely on how well you grow the bluegrasses, fescues, redtops, bents, Bermudas, St. Augustine and zoysia and on how well you can keep out a lot of pesky weeds.

In the east and north

When to plant a lawn. The only time to start a new lawn in the north is in late summer or early fall. Starting a lawn in spring is a waste of time, effort and money, because scorching sun and fast-growing weeds give the grass little chance to become well established. This does not mean, however, that if you have moved into your house in winter, you must live with bare ground all summer. On the contrary, you can get a quick coverage of green at low cost if you seed the yard with the cheapest grass seed you can buy, and then plow the grass under in the fall. To do this, you don't even have to smooth the ground first. Just scatter enough seed on it to make a carpet, and keep this watered and mowed until you're ready to make a real lawn.

Improving the soil. Grass grows pretty well in any well-drained soil, but it grows best in topsoil that is at least 6″ deep. If you're not lucky enough to have this much topsoil and if you are striving for a truly good lawn, there are three things you can do: (1) Buy topsoil and spread it over the lawn area. (2) Improve the top 6″ of whatever soil you have by mixing in four bales of peatmoss per each 1000 sq. ft. (3) Plant and plow under two or three cover crops before starting the lawn. For how to do this, see How to Improve Soil.

Preparing the soil for seeding. Whether or not you need to improve your topsoil, final preparation of the soil before starting a lawn is done in the same way:

1. Pick out large stones, roots, building scraps and other rubble.

2. Have the soil tested for acidity. If it has a pH of 5.5 or less, apply agricultural lime evenly at the rate of 50 lb. to 1000 sq. ft.

3. Apply 50 lb. of superphosphate and 20 lb. of muriate of potash per 1000 sq. ft.

4. Mix the soil thoroughly down to 6″ depth with a rotary tiller.

5. Rake smooth. Discard stones and trash.

6. To kill weed seeds in the soil as well as to supply it with nitrogen and to decompose vegetation that you have plowed under, apply calcium cyanamide at the rate of 60 lb. per 1000 sq. ft. Rake this in lightly, water well, and keep the soil damp for the next three weeks.

Seed and seeding. Whereas in the spring you should not sow anything except cheap grass seed, in the fall you should always buy and sow the best.

In the north, pure Merion Kentucky bluegrass produces a magnificent lawn if it has lots of sun. But if the lawn is partially shaded, a mixture of bluegrass and Pennlawn red fescue is recommended.

The seeding is done at least three weeks after you apply calcium cyanamide to kill weed seeds. At that time, rake the soil lightly to loosen the surface. Then scatter the seed by hand or spread it in a wheeled spreader. The package tells you how much seed to apply; but if you don't want to be bothered with arithmetic, just scatter the seed until you see almost as much of it as of the ground.

Now rake the area very lightly with the back of your rake. Don't try to bury the seed; just settle it into the soil surface. Then dampen the soil thoroughly with a gentle spray.

After-seeding care. While the seeds are germinating and until the grass is about 1″ tall, keep the soil damp. Thereafter, you can water less often, but apply more each time so that it gets down to the root tips.

Mowing should start as soon as the grass is 1½″ tall. Continue until growth stops in the fall. By then you should have a thick, green, weed-free lawn.

Early spring care. To get a lawn off to a strong start each year, there are a number of chores you must do when the frost goes out of the ground and the soil dries out enough so that you don't sink in to your boot-tops. Here are four of them. Others are discussed on the pages following:

1. Rake the grass well with a wire or bamboo rake to remove dead grass and embedded leaves.

2. Sprinkle grass seed on bare patches that may have developed.

3. Roll the lawn to flatten frost-heaved clumps. But don't use too heavy a roller, because if you compact the soil too much, the grass will stop growing. And don't attempt to level large areas because it just can't be done with a roller. The proper way to fill hollows is to spread topsoil (no more than ½″ deep) in them and sow a little grass seed.

4. Aerate the lawn. This is especially important if you have rolled the lawn or if there are spots where the soil has been packed hard by traffic. The best type of aerator to use is the power machine available at rental stores. However, small spiked rollers that are rolled back and forth by hand do a fair job.

Feeding. There are two basic rules about fertilizing lawns: (1) Give them plenty every year. (2) Use a balanced commercial fertilizer with a high nitrogen content (for example, 16-8-8 or 10-6-4).

For recommended frequency and rate of application, follow the directions on the fertilizer bag. Fertilizer with fast-acting nitrogen materials such as ammonium sulfate, ammonium nitrate, and ammonium phosphate must be applied in small doses about four times a year. Fertilizers with slow-acting nitrogen materials such as the urea-form compounds are applied in much heavier doses only twice a year.

In the north, the best time to feed a lawn (with slow-acting fertilizer) is in early spring and late August.

Lime is needed on lawns only if the soil has a pH value of 6.0 or lower. Spread it at the rate of 50 lb. per 1000 sq. ft. about once every 2–3 years. The application should be made at least a week before you apply fertilizer.

In spreading fertilizer and lime as well as grass seed and other materials, best results are obtained with a wheeled spreader, provided that either the spreader manufacturer or the fertilizer manufacturer gives directions for the setting to be used on the spreader. This varies with the fertilizer. If the information is not available, ask your garden supply store which setting is used for a comparable fertilizer. When operating the spreader, always work up and down the yard in straight lines, and avoid overlapping the strips.

If spreading of fertilizer, grass seed, etc. is done by hand, accurate distribution can be achieved only by marking off the lawn area in 10′ squares and weighing out the amount of fertilizer for each square. Scatter the fertilizer in two directions so that the ground is evenly covered.

Weeds. Theoretically, if you start with a strong, thick lawn like the one which will result from the instructions given above, you should never have to worry about weeds. But the truth is that if neighbors have weed-choked yards, you don't have a chance.

Occasional weeds that become

established in a lawn are in most cases most easily dug out with a knife. However, dandelions have such long taproots that it is better to sprinkle them with 2, 4-D or 2, 4, 5-T. This kills them within a few days.

Heavy infestations of broad-leaved weeds, such as plantain and chickweed, should also be dusted or sprayed with a mixture of 2, 4-D or 2, 4, 5-T. Crabgrass is treated as soon as it appears with DMA or PMA. But a more reliable way to stop this sprawling pest (as well as several other common weeds) is to spread a pre-emergence control, such as Tupersan or Betaran, on the lawn before the weed appears. It prevents germination of the seeds.

Other pests. Trying to trap moles or kill them with poisoned baits is an almost hopeless task. But they will disappear in a hurry if you kill the grubs on which they feed by spreading chlordane or calcium arsenate on the lawn in late winter. These chemicals also kill earthworms and sod webworms in addition to crabgrass.

Grass that is infected in warm, humid weather by a fungous disease known as brown patch should be treated immediately in this manner: First, water thoroughly. Then, 48 hours later, sprinkle or spray the affected areas with Tersan, Kromad or Acti-dione. Repeat this treatment twice more at weekly intervals. At the same time, make a practice of removing grass clippings; and if you water the lawn, water only in early morning. (Dollar spot, a somewhat similar fungous disease, is treated by spraying with Kromad or Acti-dione.)

To prevent snow mold from turning grass light brown in winter and early spring (the patches appear in poorly drained areas or where snow covers the lawn for much of the winter), do not fertilize the lawn in late fall. Keep grass cut. Apply lime if soil is acid. Spray or sprinkle grass before the first snow with mercury or cadmium-containing fungicides made for the purpose. An additional application may have to be made in the winter if the snow should melt.

Watering. Most lawns go bad because they don't get enough water, but it is possible to give them too much. A good rule of thumb is to make sure your lawn receives the equivalent of 1″ of rainfall each week. This means you have to do more than turn on the sprinkler for a few minutes at a time. When you water, soak the soil well.

Extra-heavy watering is needed where large rocks or impervious clay lie within about 8″ of the surface. In dry spells the soil in such areas dries out very quickly, and the grass dies.

Not all brown patches are caused by dryness, however. As noted above, they may also be caused by fungi which become active when the grass receives too much moisture. If you notice such spots after you have been watering the lawn zealously, stop watering for a while.

Mowing. Never cut grass too short. In the north, set your mower at 1½″ height and keep it there.

Current opinion holds that it is best to remove grass clippings from lawns, but it is actually not necessary to do so as long as the clippings are not much over ¾″ in length.

In the south.

The technique for establishing and maintaining a lawn in the north is equally good in the south—but with the following exceptions:

1. Establish the lawn in late spring or early summer.

2. Whatever else you do to improve the soil, note that almost all southern soils need more humus. To supply this, mix peatmoss into the upper 6″ of soil at the rate of 3–4 bales per 1000 sq. ft.

3. If you prefer to sow seed, you are pretty much limited in the south to the use of common Bermuda grass, carpet grass or centipede grass. These grasses have some advantages but do not produce the best lawn. Far better results are obtained with improved Bermuda grasses, zoysia or St. Augustine grass. These are planted in the form of little, growing, rooted sprigs.

After the soil has been worked, fertilized, leveled and raked, plant the sprigs in straight rows across the lawn. Space zoysia sprigs, which spread slowly, 4″–6″ apart, and leave almost all of the leaf surface above the soil (in other words, just cover the roots with soil). Space St. Augustine sprigs 8″ apart, and leave only the upper half of the leaf surface exposed above the soil. Space improved Bermuda grass sprigs 8″–12″ apart, and plant so that only the very upper ends of the leaves are exposed.

After planting the sprigs, water the lawn gently but thoroughly and keep it moist until the sprigs begin to grow.

4. As in the north, apply a balanced fertilizer that is rich in nitrogen according to the manufacturer's directions from February through October. Note, however, that improved Bermuda grasses need to be fed twice as often as zoysia and St. Augustine. In fact, in warmest climates, some gardeners fertilize Bermuda grass every month with fast-acting nitrogen fertilizer.

5. In using weed-killers on St. Augustine grass, apply at half the recommended strength.

6. Proper mowing height for Bermuda grass is ¼″; for zoysia, ½″–1″; for all others, 1½″–2″. Remove grass clippings in the south.

7. Control pests as follows: Insects—spray with malathion. Grubs, moles, armadillos, mole crickets—apply chlordane to the lawn. Brown spot and dollar spot—see instructions for the north. Gophers—set traps in the tunnels.

In the west

If you live in the cooler, more moist parts of the west, you can grow grass in the same way as gardeners in the north and east. If you live in the hot, dry southwest, grow improved Bermuda grass as in the south.

GRISELINIA One species called kupuka tree. Evergreen shrubs to 10′ with handsome, thick leaves. Grows in warm parts of California.

Plant in spring or fall in average soil. *G. littoralis* needs sun; *G. lucida* does well in light shade. Water well and often. Fertilize every 2–3 years.

GROUND CHERRY *Physalis pubescens*, husk tomato. Annual to 30″ with round, yellow, cherry-size, edible fruits in papery husks. Grows almost everywhere.

Grow like Chinese lantern (which see).

GUATEMALA HOLLY *Olmediellia betscheleriana*. Evergreen tree to 25 with holly-like foliage. Grows in warmest climates.

Plant in spring or fall in average soil in the garden or a tub. Grow in sun or partial shade. Water in dry weather. Fertilize every 3–4 years in spring. Prune in winter to direct and control shape.

GUAVA *Psidium guajava.* Evergreen shrubs to 30′ with purple or yellow fruits used for jelly. Grows in subtropics.

Plant in spring in the sun in average, well-drained soil which contains a fair amount of humus and some well-rotted or dried manure. Water in dry weather. Fertilize every 6–8 weeks. Thin fruits when they are set in order to force remaining fruits to grow larger. Prune in winter. If tips of new stems are kept pinched back, plants will be bushier and lower.

GUINEA GOLD VINE *Hibbertia volubilis.* Dense evergreen vine to 30′ with dark-green foliage and yellow flowers in summer. Subtropical.

Propagate by stem cuttings. Plant in well-drained, sandy soil containing humus. Grow in partial shade. Keep soil moist. Fertilize every spring. Provide sturdy support for vine to twine up on. In early spring, cut back to about 20′ and thin out excess growth.

GUNNERA Evergreens to 6′, resembling gargantuan rhubarb, with huge leaves and 3′ spikes of colorless flowers. Grows in warmest climates.

Propagate by division of roots. Plant in spring in filtered sun in a location not exposed to much wind. The soil must be very rich in humus and always kept moist. Fertilize in spring and summer with general-purpose plant food with a high nitrogen content, or use nitrate of soda. Spray foliage with water frequently. Cut to the ground in winter.

HABENARIA Fringed orchis. Perennial wildflower to 3′ with beautiful clusters of flowers in various colors. Grows almost everywhere in cool, moist bogs.

If you are very careful, you may be able to dig plants from the wild 6–8 weeks after they have bloomed. Take as much soil as possible. Plant in partial sun. Since habenarias are bog plants, they require very acid, moist, humusy soil. Protect from slugs, snails and mice. Be warned, however, that these are difficult plants to move and to raise successfully; you probably will be doing yourself and the world a favor to leave them in their natural habitat.

HACKBERRY *Celtis,* sugarberry, nettle tree. Deciduous trees to 100′, somewhat like elms. Grows mainly east of the Rockies.

Plant in early spring or fall in sun in average soil. Water regularly until established, thereafter in dry weather. Fertilize every 3–5 years.

HAKEA One species called sea urchin. Evergreen shrubs to 30′ with narrow or needle-like leaves and showy clusters of white or red flowers. Grows in warm parts of California.

Plant in spring or fall in sun in well-drained, sandy, average soil. Once established, water only in summer and then not too much. Fertilize every 2–3 years in early fall.

HAWORTHIA Succulents to 6″, some forming rosettes, others columnar, others with windows in the leaves. Grows indoors; also outdoors in hot, dry climates.

For how to grow, see cactus, desert. Give light shade in summer. Repot house plants every fall.

HAWTHORN *Crataegus*, thorn, thornapple, haw. Deciduous trees or shrubs to 30′ with wicked thorns, fragrant white or pink flowers in spring and small, apple-like fruits in fall. Grows mainly in the east and south.

Plant in early spring in sun or partial shade. The soil need be of only average quality, well drained, preferably sandy and slightly alkaline. Water in dry weather. Fertilize every 2–3 years. Prune after flowering. If cankers develop on branches, cut out the branches entirely and burn.

HAY-SCENTED FERN *Dennstaedtia punctilobula*, boulder fern. Graceful, light green fern to 2′ with sweetly scented fronds. Grows mainly in the northeast quarter of the country.

For how to grow, see fern. Grow in semi-shade or sun in average soil.

HAZELNUT *Corylus*, hazei, filbert. Deciduous shrubs and trees to 60′ with small, edible nuts and some with bright autumn foliage. Grows in all but most extreme climates. See also filbert.

Plant common hazelnuts in spring or fall in sun or light shade in average soil. Water in dry spells. Fertilize every 3–4 years.

HEART SEED *Cardiospermum halicacabum*, balloon vine. Slender vine to 10′ with tiny, balloon-like pods containing black seeds with a heart-shaped, white spot. Needs a long growing season to mature, is therefore best in warm climates.

Sow seeds when frost danger is past in light, well-drained loam improved with fertilizer. Grow in sun and do not expose to too much wind. Keep watered. Fertilize lightly every month during summer. Provide a light trellis for vine to climb on.

HEATH *Erica*. Evergreen plants to 18′ but usually much lower, with needle-like foliage and white, red or pink flowers usually in small spikes. Grows in mild and warm climates where the summers are not too hot.

Grow like heather (which see). Stem cuttings of heath should be made in winter or early spring. Cut back plants to the base after flowering to promote bushiness.

HEATHER *Calluna vulgaris*. Evergreen shrubs to 3′ with small leaves and dense flower spikes of white, pink, red or yellow. Grows best in cool climates.

Propagate by stem cuttings of young wood. Plant in spring or fall in the sun. The soil should be a well-drained mixture of one part loam, one part sand and one part peatmoss. Mix in a little dried manure. Also mix in aluminum sulfate if soil is not acid: heather requires considerable acidity. During their first winter, new plants should be covered lightly with straw. Water well in dry weather. Shear the plants lightly in early spring.

HEBE Also identified as *veronica*. Evergreen shrubs to 6′, inclined to be somewhat straggly, with flower spikes of purple, blue, white or rose. Mainly grown in California.

Plant in spring or fall in rich, moist, humusy soil. Grow in sun in milder areas, in partial shade in hot climates. Water regularly while making growth. Fertilize every 1–2 years in spring. After bloom, cut flower-bearing stems to the ground.

HEDGEHOG CACTUS *Echinocereus*. Different species called lace cactus, rainbow cactus. Cylindrical or round desert cacti to 1′ with vicious spines and bright flowers in many colors. Grows indoors; also outdoors in hot, dry climates.

For how to grow, see cactus, desert. Water moderately even when growing. Flowering is improved if plants are not exposed to below-freezing temperatures, though the plants themselves are not harmed except by very low readings.

HEDYCHIUM Butterfly lily, ginger lily. Rhizomatous plants to 6′ with fine foliage and beautiful flowers often white or yellow and red. Grows outdoors in warmest climates.

Grow like *alpinia* (which see).

HELENIUM Sneezeweed, yellow star, false sunflower, Helen's flower. Perennials to 6′ with clusters of usually yellow, ray flowers in late summer and fall. Grows almost everywhere.

Divide plants every spring. Plant in a sunny location in average soil with extra humus to hold moisture. Water often. Fertilize once or twice. Pinch back growing tips for the first 1–2 months to promote bushier growth.

HELIANTHEMUM Sun rose, frostweed. Evergreen perennial to 2′ with fleeting, yellow flowers. Grows in the east and California.

Propagate by spring division. Plant in sun in sandy, well-drained soil that is limed occasionally. Water sparingly. Feed in spring.

HELICONIA Wild plantain, false bird of paradise, lobster claw. Perennials to 12′ with very showy, red, pink or orange bracts. Grows in warmest climates.

Divide in fall. Plant in sun or partial shade in a place where plants will have plenty of space. Soil should be rich, humusy and well drained. Water regularly except in winter, when supply should be reduced. Fertilize 2–3 times while plants are making growth in late winter, spring and summer.

HELIOCEREUS One species known as sun cereus. Slender, sprawling desert cacti to 3′ with beautiful red flowers. Grows indoors; also in hot, dry climates.

For how to grow, see cactus, desert. Don't let temperature drop below 50°.

HELIOPSIS Orange sunflower, false sunflower. Perennial much like the sunflower but to only 5′. Native in the east and southeast, but grows almost everywhere.

Divide in spring and plant in a sunny spot. The soil need be only of average quality but with extra humus. Water in dry weather. Fertilize in spring. Divide plants every three years. Spray with malathion to control aphids.

HELIOTROPE *Heliotropium*. Perennial best grown as an annual, to 4′, with fragrant, terminal clusters of flowers in violet or white. Grows best in warm climates. Also a house plant.

You can have heliotrope the year round by growing it in the following way: About January or February take cuttings from plants growing indoors. Plant rooted plants outdoors in sun or very light shade when danger of frost is past. Heliotrope needs a well-drained, humusy soil which is kept moist and is fertilized monthly. In June or July, take cuttings from the garden plants and propagate for winter bloom. As soon as they are rooted, pot in equal parts of loam, humus, and sand. Shift into larger pots as needed (heliotrope needs root room). Keep in light shade outdoors and pinch any buds. Move indoors before frost and place in a sunny, very cool window. Keep soil moist and spray the foliage with water often. Fertilize monthly. Spray with malathion to control white flies.

HELIPTERUM An annual everlasting to 20″ with rose-pink flowers. Grows almost everywhere.

Sow seeds outdoors after danger of frost is past, or indoors about eight weeks before last frost. Grow in full sun in average soil enriched with a little fertilizer when young plants are established. Water in dry weather. To dry, cut flowers before they are wide open, strip off leaves and hang flowers upside down in a shady place.

HELLEBORUS Christmas rose, winter rose, Lenten rose, hellebore. Evergreen perennial to 1′ with white or pinkish flowers in winter. Grows in temperate climates.

In spring divide roots so that each section has several eyes. Plant these 1″ below the soil surface in a spot which is shaded in summer, sunny in winter. The soil should be well drained, sweet, rich in humus and fertilized several times with plant food rich in phosphorus. Water regularly in summer. Mulch soil in summer with peatmoss to hold in moisture and keep roots cool. Do not disturb plants once they are established.

HEMLOCK *Tsuga.* Graceful evergreen tree to 120′ with horizontal or pendulous branches and short, flat needles. One variety is weeping. Grows best in the northeast, the Appalachians and the northwest.

Plant in spring or early fall in partial shade or sun. Do not expose to wind and do not attempt to grow in a smoky city or industrial atmosphere. Grow in average, well-drained soil containing plenty of humus. Water in drought. Fertilize for the first few years. Canadian hemlock will develop more as a bush with several trunks if the secondary trunks are not removed as they develop. To control size and shape, prune every spring, though this destroys their lovely feathery character. Spray with malathion if infested with red spider.

HEPATICA Liverleaf, liverwort. Perennial wildflowers to 6″ with small, blue, white, purple or pinkish flowers in early spring. Grows best in cool regions east of the Rockies.

Divide plants after flowering. Plant in light shade in average soil that contains some extra humus but is somewhat gravelly and well-drained. Don't worry that plants die down after flowering; they will soon produce a new set of leaves that will last until the next spring.

HERALD'S TRUMPET *Beaumontia grandiflora,* Easter lily vine. Woody vine to 30′, with large, dark-green leaves and hundreds of large, fragrant, white flowers like Easter lilies. Subtropical.

Propagate by stem cuttings and plant in deep, well-drained, rich, humusy soil in full sun in spring or fall. Protect from wind. Water in dry weather. Fertilize in early spring. Guide the stems upward on a strong trellis. Right after flowering, cut back hard (but keep some of the old wood on which next season's flowers will develop).

HEUCHERA Coral bells, alumroot. Perennials to 28″ bearing dainty clusters of small, bell-shaped, white, red or purple flowers. Coral bells grows in all but most severe climates. Other species are eastern or western plants.

Divide roots or sow seeds in spring. Grow in average, well-drained soil with extra humus. Needs sun or partial shade. Water regularly. Fertilize in early spring and once again during the growing season. Topdress in spring with humus. Spray with malathion to control mealybugs.

HIBA ARBOVITAE *Thujopsis dolabrata,* false arbovitae. Evergreen tree to 50′ very closely resembling arborvitae. Grows in mild, moist climates.

Grow like arborvitae (which see).

HIBISCUS MOSCHEUTOS Rose mallow, swamp mallow, sea hollyhock, marshmallow. Perennial to 7′ with large, hollyhock-like flowers in white and reds. Grows in temperate climates. See also Chinese hibiscus and rose of sharon.

Divide in spring and plant in sun in rich soil. Water regularly in dry weather. Fertilize in spring.

HICKORY *Carya.* Deciduous trees to 130′, prized for their wood and nuts. Grows mainly from Iowa and Texas east. See also pecan.

Plant in spring or fall in a sunny spot in average, well-drained, moist soil. Cut back young trees one-third. Water for the first 1–2 years. Fertilize every 3–5 years. If trees are grown for their nuts, more care is required (see pecan).

HOFFMANNIA Plants to 4′ with handsomely colored foliage and unimportant red or yellow flowers. A house plant.

Propagate by stem cuttings in spring. Pot in two parts loam, two parts humus, and one part sand. Keep moist. Grow in a warm east or west window. Stand pots on wet pebbles and occasionally spray foliage with water. Fertilize 3–4 times a year.

HOLLY *Ilex.* Handsome evergreen and deciduous trees and shrubs to 40′ with glossy foliage and red or black berries. Generally grows best in moist, temperate climates. Berries are produced only if male and female plants are planted together.

Plant in spring or fall in sun or extremely light shade. Grow in a light, well-drained soil containing humus. The soil can be neutral or slightly acid, but not very acid. Water frequently and deeply for several weeks after plants are set out; thereafter, water when weather stays dry for some time. Fertilize annually in the spring. Prune in early spring before growth starts.

Screen plants that are exposed to cold winter winds.

HOLLY FERN *Cyrtomium falcatum.* Stiff fern to 2′ with shaggy stalks and glossy leaves something like a holly leaf. A house plant; also grows outdoors in warmest climates.

For how to grow, see fern. Holly fern does not resent house conditions as much as many ferns, but prefers to be in a cool room at night during winter.

HOLLYHOCK *Althaea rosea.* Biennials to 9′ with single or double summer flowers in many colors. Grows almost everywhere.

Sow seeds outdoors any time until the end of July, and move plants to their location in a sunny border the next spring. Plant a little deeper than they formerly grew. Grow in average, well-drained soil. Water regularly. Fertilize in spring. Spray with zineb every week to prevent rust. Stake plants exposed to wind. Burn plants in the fall. Hollyhock reseeds itself with abandon, but these plants eventually revert to type.

Hollyhocks may also be treated as annuals if seeds are sown indoors 8–10 weeks before last frost.

HOLLYLEAF SWEETSPIRE *Itea olicifolia.* Lovely evergreen shrub to 10′ with glossy, holly-like leaves and pendulous, white, spring flower spikes. Grows in warmest parts of California.

Plant in early spring or fall in partial shade in average soil. Water regularly during summer. Fertilize every 2–3 years in spring. Prune after flowering until Christmas time to control shape.

HOLODISCUS DISCOLOR Cream bush, ocean spray. Deciduous shrubs to 20′ with hairy, greyish leaves and clusters of creamy flowers. Grows west of the Rockies.

Propagate by layering. Plant in spring or fall in sun or partial shade. The soil should be well-drained, rich and contain considerable sand and humus. Water in dry weather. Fertilize every 2–3 years in spring. Prune after flowering.

HONESTY *Lunaria,* money plant, moonwort, satinpod, satinflower. A biennial usually grown as an annual. A 30″ everlasting with flat, silvery, parchment-like parts in fall. Grows almost everywhere.

Sow seeds indoors about eight weeks before last frost and shift into a partially shaded location in average soil in the garden. Water in dry weather. Fertilize when young plants are established. Cut stalks after money-like parts are round, and dry them in a shady place. If honesty is grown as a biennial, start seeds in a seedbed in June or July and transplant to their permanent positions in the fall.

HONEY BUSH *Melianthus major.* Evergreen shrub to 12′ with leaflets arranged in feather shape and reddish-brown flower spikes in late winter. Grows in warm parts of California.

Plant in spring or fall in sun or light shade in average soil. Water in unusual dry spells; otherwise plant needs little attention. Cut out old stems annually when plant starts making growth.

HONEYDEW MELON *Cucumis melo.* Melon with smooth, ivory-colored rind and green flesh. Requires 110 days to mature and is therefore best grown in warm climates.

Grow like cantaloupe (which see).

HONEY LOCUST *Gleditsia.* Deciduous trees to 100′ with light, lacy foliage. Grows in east, south and midwest. See also locust.

Honey locust tolerates city conditions, needs little attention. Plant in early spring or fall in sun in average soil. Water and fertilize while tree is getting started. Thereafter, water only in drought. Feed every 3–5 years.

HONEYSUCKLE *Lonicera.* Large genus of woody vines and shrubs to 30′, some evergreen, most deciduous, with flowers in many pastel shades and fruits in various colors. Different species and varieties grow in different climates.

Propagate by cuttings of young stems or by layering. Plant in average, well-drained soil which should be kept moist. Fertilize every 1–2 years in the spring. Honeysuckles prefer sun but do well in partial shade. Provide a sturdy support for climbing trees. Prune in early spring or fall to remove dead wood and control growth.

HOP HORNBEAM *Ostrya virginiana,* ironwood, leverwood. Deciduous tree to 30′ with open head and bright-green foliage Widely distributed through eastern half of the country.

Plant in spring or fall in average soil in sun. Needs little attention or water Growth is slow.

HOREHOUND *Marrubium vulgare,* hoarhound. Perennial herb to 3′ with grey-green, aromatic leaves used for flavoring medicines and candy. Grows in mild and warm climates.

Divide plants in spring or fall and plant in sun in almost any soil. Water in very dry weather.

HORNBEAM *Carpinus,* blue beech, ironwood, water beech. Shapely deciduous tree to 50′ with fine foliage. Grows in temperate climates.

Plant in early spring or fall in sun or partial shade. Hornbeams grow in almost any soil, need little attention. Give some protection from winds. The European hornbeam may be grown as a tall hedge and can be pruned severely in winter.

HORSE CHESTNUT *Aesculus,* buckeye. Deciduous tree to 100′ with showy white, pink or red spring flowers and spiny fruits containing one or two large, shiny seeds. Grows in mild and warm climates.

Plant in early spring or fall in a sunny location in average soil. Water in dry weather. Fertilize every 3–4 years; more often if soil is on the poor side.

HORSE-RADISH *Armoracia rusticana.* Plant to 30″ grown for its pungent, white root. Grows almost everywhere.

Horse-radish grows best in rich loam, but tolerates all soils except those that are very light and sandy. Dig the soil deeply and pulverize well in the spring. Plant the root cuttings large-end up in trenches and cover with 2″ of soil. Space the cuttings 10″ apart in rows 2′ apart. Keep cultivated and watered during the summer. Dig up roots in the fall when the weather turns cold and grate immediately. However, roots can be left in the ground all winter. The following spring plant new root cuttings since roots over one year old deteriorate. Cuttings are made by slicing the roots lengthwise into ¼″ strips about 6″ long. Plant at once.

HOSTA Funkia, plantain lily. Perennials to 3′ grown primarily for their handsome leaves, but with pleasant little, blue, lavender or white flowers in summer. Grows in all but extreme climates.

Divide young plants in the spring and plant in average, well-drained soil to which humus has been added. Fertilize once or twice in spring and summer with nitrogen-rich plant food. Keep fairly moist. Cultivate regularly and mulch in winter. By and large, hosta does best in light shade, but also does pretty well in sun. Plants with blue-green foliage achieve best color in sun.

HOWEA Different species called Belmore sentry palm and Forster sentry palm. Sometimes identified as *kentia*. Slender palm trees to 30′ with ringed trunks crowned with a graceful cluster of slightly drooping fronds. Grows indoors; also in subtropics.

For how to grow, see palm. The Forster sentry palm should be grown in partial shade. Protect both species against dry winds and cold.

HUCKLEBERRY *Gaylussacia*. Different species called dangleberry, buckberry, juniperberry, gopherberry. Evergreen and deciduous shrubs to 6′ with blue or black berries closely resembling blueberries. Grows mainly east of the Mississippi.

Grow like blueberry (which see), but plant in light shade.

HYACINTH *Hyacinthus*. Spring bulbs to 15″ with large, fragrant, stiff flower clusters in various lovely colors. Grows almost everywhere; also indoors.

Outdoors. Plant in the fall where flowers will get at least four hours of sun a day. The soil should be well drained, dug to 1′ depth and improved with humus. Set the bulbs 6″ deep and 6″ apart, and water well. Mulch the soil after it freezes with salt hay or leaves. This is required only in the first year. In early spring and again in the fall, scratch a little general-purpose plant food into the soil. After flowering, cut off flower stalks but don't touch the leaves until they die down completely; then gently tug them loose.

After a number of years, hyacinth flower clusters tend to become sparse and open. Replace bulbs at this time.

Indoors. See How to Force Bulbs for Indoor Bloom. Hyacinths can also be grown indoors in water. To

do this, fill special hour-glass-shaped hyacinth glasses or any narrow jars with water so that it almost touches the bottom of the bulbs set in the top of the glasses. Keep in a dark, 50° place for 10–12 weeks while roots develop. When the buds are well out of the bulbs, gradually move the plants into a warmer, brighter place. After about 10–14 days, move into a sunny window to bloom. Discard bulbs when flowers die.

HYACINTH BEAN *Dolichos lablab.* Ten-foot perennial vine treated as an annual. It is covered with long-lasting, purple or white flowers in summer. Grows in warm climates.

After danger of frost is past, sow seeds where they are to grow in average, fertile soil in a sunny, sheltered location. Keep watered. Fertilize when seedlings are thinned out and monthly thereafter. Grow vine on a stout trellis. Pull up in the fall when it dies down.

HYDRANGEA Deciduous shrubs to 15′ with great white, blue or pink flower clusters. Grows almost everywhere. Also a house plant. See also climbing hydrangea.

Outdoors. Propagate by half-ripe stem cuttings or by layering. Plant in spring or fall in good soil containing a substantial amount of humus and some well-rotted or dried manure. The plants can grow in sun or light shade except in warm climates, where they should be grown on the north side of the house or in a spot where they get only a few hours of sun. Water regularly while making growth, especially in dry weather. In dry climates, spray the foliage with water occasionally. Fertilize in spring. *H. paniculata* should be pruned rather severely in the spring, otherwise it may overpower you. Other popular hydrangeas are pruned right after flowering.

To change the color of French hydrangea (*H. macrophylla*) flowers from pink to blue, dose each plant with a gallon of water in which 3 oz. of aluminum sulfate have been dissolved. Make the first application as soon as frost is out of the ground and make 4–6 additional applications at ten-day intervals afterwards. To turn French hydrangea flowers pink, apply a cupful of superphosphate to the soil and water it in well.

Indoors. If you are given a hydrangea, grow it in a cool north or east window. Water thoroughly once or twice a day. When flowering stops, cut it back 50 per cent, repot in fresh general-purpose potting soil, reduce watering slightly and then, when weather warms, plunge the pot outdoors in partial shade. Water regularly and fertilize monthly. Then, just before frost, bring the plant into a very cool, dim place and water very little until January, when the plant should be moved into a somewhat brighter place at about 50°. Water a little more heavily. When growth starts, move into a south window at about 60°. Water daily and fertilize every month.

An alternative to the above procedure is to take stem cuttings from your gift plant in late winter or spring. When rooted, pot up in general-purpose potting soil and grow in a sunny window with plenty of water. Move outdoors after frost. Proceed from there on as above.

HYSSOP *Hyssopus officinalis.* Perennial herb to 18″ with red, white or blue flowers. Grows in temperate climates.

Divide in spring and plant in sun in average, well-drained soil which should be limed occasionally. Water in dry weather. Fertilize in spring.

INCENSE CEDAR *Libocedrus decurrens.* Handsome evergreen to 90′ with narrow, pyramidal crown and scale-like leaves. Grows in mild climates, but best on the West Coast.

Plant in spring or fall in partial shade in deep, well-drained soil. Water deeply once a month in dry weather. Fertilize every year or two while tree is young and getting started.

INULA Sunray flower, elecampane, horseheal. Perennial to 6′ with daisy-like, yellow flowers. Grows in temperate climates.

Divide in spring or fall and plant in average soil in a sunny location. Water sparingly. Fertilize in early spring. Stake tall plants.

IRIS Flag. Large genus of perennials growing from rhizomes or bulbs to a height of 42″, with flowers in an incredible array of colors. Grows almost everywhere.

Bearded iris (including the German iris, I. pumila and innumerable named hybrids). Plant the rhizomes in sun in deeply dug soil that is well drained and contains some peatmoss and a little fertilizer. Plant in July and August except in hot, dry climates, where early fall is better. Space the rhizomes of large varieties 9″–12″ (or even up to 18″, if you're not looking for a quick clump) apart; space smaller varieties about 6″ apart. Just cover the rhizomes with soil and water well until growth starts. From then on, water only when soil becomes very dry.

Fertilize every spring as growth starts. Keep dead flower stalks cut and remove dead foliage, but don't bob the tops of the leaves except when dividing clumps. Cultivate carefully and very shallowly. Spray with sevin and zineb every two weeks to control borers and disease. If rhizomes develop a soft rot, cut out wound and treat as for borers (see below). Divide clumps every 3–4 years or when the rhizomes become crowded. To do this, cut out and discard old sections of the root. Save the new plump sections. Each should include a small fan of leaves, which should be trimmed to a height of 6″. If you find borers in the rhizomes, soak in corrosive sublimate (deadly poison) or in a quart of water to which 2 tsp. of Semesan have been added.

Beardless iris (including Siberian iris, Japanese or kaempfer iris, Louisiana iris, I. spuria). Propagate by cutting the small, much-rooted rhizomes into sections containing several sets of leaves. Do not keep the roots out of the ground any longer than possible. Just cover with soil. Space the rhizomes about 1′ apart (Siberian iris rhizomes, however, are often planted only 3″–6″ apart). The soil required by beardless irises should be rich, very humusy and somewhat acid. The time of dividing and planting is as follows: Louisiana iris, right after flowering; Siberian iris and *I. spuria,* September; Japanese iris, July–September.

Beardless iris grows in sun as a

rule, but the Louisiana type prefers filtered sun. Water regularly while plants are making growth and flowering. The Japanese and Louisiana irises are particularly thirsty types. Fertilize in early spring and once more in late spring (or dress plants in summer with well-rotted manure instead of feeding them a second time with balanced plant food). Cultivate shallowly. Remove dead flowers and leaves but don't cut foliage except when dividing plants every 3–5 years. Mulch the soil around newly set out plants after it freezes in the fall. Treat pests in the same way as on bearded iris.

Bulbous iris (including I. reticulata, Dutch iris, Spanish iris, English iris). Plant bulbs 4″ deep in fall in average, well-drained soil in a sunny location. Water regularly in spring until flowering stops, then gradually reduce the supply. Fertilize in the spring. Allow foliage to die down naturally; do not cut. Lift and divide bulbs when clumps become crowded after foliage dies down. In cold climates, mulch in winter with leaves or evergreen boughs.

Crested iris (including I. cristata, I. gracilipes, I. tectoroum and I. japonica). Divide rhizomes in summer (right after flowering in the case of *I. cristata*). Plant in light, well-drained soil containing a fair amount of humus. Give *I. japonica* (which grows only in warm climates) shade. Others want sun in the morning but not all day long. Water in dry weather. Fertilize in spring.

Iris indoors. See How to Force Bulbs for Indoor Bloom. The varieties used for forcing are *I. reticulata* and the Wedgewood variety of Dutch iris. Plant 1″ deep in early October or at any time thereafter.

IRISH HEATH *Daboecia cantabrica*. Evergreen shrub to 18″ with terminal clusters of purple or white flowers in summer and fall. Grows in warm and mild climates.

Grow like heather (which see).

ISLAND BUSH POPPY *Dendromecon rigida harfordi*. Evergreen shrub to 20′ with greyish-green leaves and yellow, poppy-like flowers. Grows in dry parts of California.

Plant in spring or fall in sun in average soil with excellent drainage. Water deeply about once a month in dry weather, especially when plant is young. Fertilize very occasionally. Cut back hard in late winter.

IVY *Hedera*. Vigorous, handsome evergreen vines to 90′ with scores of varieties. Grows almost everywhere, except in most severe climates, if proper varieties are used. Also grows indoors.

Outdoors. Propagate by layering or by tip cuttings started in water. Ivy is not overly particular about soil but does best in fertile, well-drained loam. Grow in partial shade in cool regions, full shade in warm regions. Water regularly and spray foliage occasionally with water during warm weather. Fertilize every 2–3 years. Provide a fairly rough surface for aerial rootlets to cling to. In early spring, prune to control growth, and thin out vine if it is too dense. Spray with malathion if attacked by insects.

Indoors. Pot in general-purpose potting soil which should be kept moist but not wet. Fertilize lightly every month. Keep in a north window at a temperature no higher than 60°. Spray foliage with water every week, or place under a faucet. Pinch stem ends often to encourage bushing out.

Ivy can also be grown in a jar of water to which a lump of charcoal is added. It needs same treatment as above.

IXIA Corn lily. Plants growing from corms to 2′, with hanging, bell-shaped flowers in many colors. Grows outdoors in warm climates. Also a house plant.

Outdoors. Plant in fall in a sunny, protected spot. The corms go 2″ deep and 4″ apart. The soil should be well-drained, of average quality. In areas where the ground freezes, mulch in winter. Still further north, ixias can be grown like gladiolus (which see).

Indoors. Pot in the fall in general-purpose potting soil. Set in a very cool, sunny window. Grow like freesia (which see).

IXIOLIRION Siberian bluebell. Bulbous plant to 1′ with a cluster of blue flowers in spring. Grows in temperate climates.

Plant bulbs 3″ deep and 4″ apart in average, well-drained soil in the fall. Give full sun. In very cold climates, bulbs should be dug up in the fall, stored indoors in a cool, dry place and planted out again in early spring.

IXORA Evergreen shrubs to 6′ with brilliant flower clusters in reds, oranges and yellows much of the year. Grows outdoors in warmest climates. *I. coccinea*, called flame-of-the-woods, is one of the most popular species and is often grown indoors.

Outdoors. Propagate by stem cuttings in spring. Plant in spring or fall in light shade. The soil should be a well-drained mixture of one part loam, one part coarse humus and one part sand. Water regularly. Fertilize 2–3 times a year. Prune after a spurt of flowering to control and direct growth. Flame-of-the-woods can be sheared like a hedge without affecting the flowering. Spray with malathion to control aphids, and with an oil emulsion in spring to control scale. Apply nemagon to the soil if plant is attacked by nematodes.

Ixoras also make good pot plants. If in large pots, they do not need to be repotted more than once every 2–3 years provided that you feed them liberally.

Indoors. Pot in general-purpose potting soil. Keep moist. Feed monthly while plant is making strong growth. Grow in an east or west window at not less than 65°. Give as much humidity as possible. Move outdoors into light shade in summer. To force repeat flowering, reduce water supply sharply for a month after flowers die. Then trim branches back to one joint and start regular watering and feeding.

JACARANDA Deciduous tree to 40′ with feathery foliage which is shed in early spring and blue flower clusters in spring and summer. Grows in subtropics.

Plant in spring in a sunny, not too windy location. Grow in average soil. Fertilize no more than every 2–3 years. Water deeply but not often in dry weather. If hit by frost, plant can be pruned back. Don't expect flowers until plant is a number of years old and well developed.

JACK IN THE PULPIT *Arisaema triphyllum*, Indian turnip. Tuberous-rooted, perennial wildflower about 2′ tall with a compact "flower" resembling a preacher in a covered pulpit. Grows in the east.

Transplant from the wild in early fall. Plant in shade in moist, humusy, acid soil.

JACOBEAN LILY *Sprekelia formosissima*, Aztec lily, St. James's lily, Mexican fire lily. Bulbous plant 1′ tall with orchid-like, red flowers in spring and summer. Grows in warm climates. Also a house plant.

Grow like amaryllis (which see). But note that indoors these plants need considerable humidity in their early stages. To supply this, cover each plant with a large mason jar until the flower stalk is 6″–8″ high.

JACOBINEA CARNEA King's crown. Often identified as *justicia*. Plant to 4′ with pink flower spikes. Subtropical. Also a house plant.

Propagate by stem cuttings after flowers fade in the fall. Plant in well-drained, average soil. Grow in partial shade or in an east window indoors. Water regularly in spring when plant is making growth. Fertilize garden plants once or twice at this time; fertilize house plants every six weeks during spring and summer. Give house plants as much humidity as possible, and keep warm.

JAPANESE HOP *Humulus japonicus*. Luxuriant vine to 35′, excellent for providing a quick screen or sunshade. Grows best in warm climates.

Sow seeds where they are to grow after danger of frost is past. Grow in average soil. Keep moist. Feed 2–3 times during the spring and summer. Grow on a strong trellis. Tear out in the fall.

JAPANESE PAGODA TREE *Sophora japonica*, scholar tree. Spreading deciduous tree to 60′ with feathery leaves and loose clusters of creamy flowers in summer. Grows in warm and mild climates.

Plant in early spring or fall in sun in average soil. Water in dry weather. Fertilize every 1–2 years when tree is young, thereafter, every 3–4 years.

JAPANESE RAISIN TREE *Hovenia dulcis*. Deciduous tree to 30′ with handsome foliage and small, round fruits. Grows in mild climates.

Plant in early spring or fall in sun. Grow in average soil containing a large amount of sand. Water in dry spells. Fertilize every 2–3 years.

JASMINE *Jasminum*, jessamine. Evergreen and deciduous, erect and climbing shrubs to about 20′, with white or fragrant yellow flowers at different times of year. Grows in warm climates. Some species are also good house plants. See also *gardenia* (cape jasmine).

Outdoors. Propagate by layering. Plant in spring or fall in a sunny location in good, humusy soil. Water regularly. Fertilize in late winter or early spring and again about June. Train to a strong support. Prune hard in winter. The main job is to cut back or totally remove old stems. To control size and promote bushy habit, young wood must also be trimmed back.

Indoors. Pot in general-purpose potting soil. Keep moist. Spray the plant every day with water while it is making growth. During the same period, fertilize every three weeks. Grow in a sunny window at 65°. Move outdoors in summer.

JATROPHA One species called coral plant; another, bellyache bush. Shrubs to 10′ with excellent foliage and small flowers usually red. The plant juices and seeds are poisonous. Grows in warmest climates.

Sow seeds in spring in pots or flats. Put plants in garden in spring or fall in average soil in sun or light shade. Water in dry weather. Fertilize every 1–2 years in the spring. Prune in late winter.

JERUSALEM ARTICHOKE *Helianthus tuberosus.* Perennial to 12′ with large sunflowers and edible, potato-like tubers tasting like globe artichokes. Grows best in cool climates.

Jerusalem artichoke does well in poor, gravelly soil; it develops too much top growth in anything better. Plant small tubers 3″ deep and 15″ apart in rows 30″–36″ apart. Large tubers can also be cut in half and planted the same way. Water in dry weather. No further care is required. Dig up tubers any time from fall into spring. Tubers left in the ground will come up the following spring.

Note. Jerusalem artichoke is a weed and will take over your garden in short order if most of the tubers are not dug up every year.

JERUSALEM CHERRY *Solanum pseudo-capsicum,* Christmas cherry. Shrub to 4′ with long-lasting, poisonous red or yellow winter fruits. A house plant.

Sow seeds indoors in early spring, shift into pots, move outdoors in summer, pinch stems when 3″ tall, fertilize monthly and bring in before frost. Pot in general-purpose potting soil and, after about a week in a north window, place in a south window at no more than 60°, preferably lower. Water when soil feels dry. Spray foliage with water frequently. After plant loses fruit and leaves in the winter, cut back all branches to two buds. Place in a very cool north window and water sparingly until growth begins again. Then repot in a little fresh soil and return to a sunny window as before. Move outdoors and feed as above during summer.

JERUSALEM SAGE *Phlomis fruticosa.* Shrub-like plant to 4′ covered with hairs and with yellow flowers in whorls. Grows in mild and warm climates.

Propagate by spring cuttings. Plant in spring or fall in sun in average soil. Protect against very cold weather. The plant is quite drought resistant and needs little care.

JEWEL VINE *Derris scandens.* Dense evergreen vine to 35′ with small, fragrant, white flowers in summer and fall. Grows in warmest climates.

Propagate by stem cuttings. Plant in average soil rich in humus. Grow in sun or partial shade. Don't expose to too much wind. Keep soil moist. Fertilize 2–3 times during spring and summer. Train on a post or trellis. Prune in winter.

JOB'S TEARS *Coix lacryma-jobi.* Annual grass to 6′, grown for its sprays of grey seeds, which are dried and sometimes used as beads. Grows best in warm climates.

Sow seeds in average soil in full sun in the spring. Fertilize lightly when seedlings are thinned to stand 1′ apart. Water in dry weather. Cut before the bead structure opens, and dry in a shady place.

JUJUBE *Zizyphus.* Shrubs or trees to 30′ with fine deciduous or evergreen foliage, prickly branches and edible, oblong fruits in late fall. Grows in warmest climates.

Plant in spring or fall in average, well-drained soil in a sunny location. Water in dry weather. Fertilize every spring. Keep cultivated.

JUNIPER *Juniperus.* Several species are called red cedar. A large genus of evergreen trees and shrubs, some tall (to 100′) and columnar, others low and spreading or creeping. Grows almost everywhere.

Plant in spring or early fall in full sun. Only average soil is needed, but it must be well drained. Once established, water is needed only in long dry spells. A yearly application of fertilizer will speed the growth of young plants, but feed established plants only every 3–4 years. Prune spreading types in early spring to control growth and keep them from getting too large. Spray in late spring with malathion to control juniper scale and spider mites. Sometimes orange, gelatinous growths appear on red cedars. These do little harm to the cedars but serious damage to apples, crabs, and hawthorns growing in the area; consequently, you must choose which you want—the cedars or the apples—and remove the other.

KAFIR LILY *Schizostylis coccinea.* Perennial to 2′ with tuber-like roots and a red flower spike in late fall. Grows in warm climates.

Divide roots in spring, keeping about three or more eyes to a section. Plant 3″ deep and about 1′ apart in spring or fall in the sun. Grow in average soil and water regularly. Fertilize once or twice during the summer.

KAFIR PLUM *Harpephyllum caffrum.* Spreading evergreen tree to 35′ with narrow leaflets which start out red and turn green, small white flowers, and edible, red, olive-like fruits. Grows in warmest climates.

Plant in spring or fall in sun in average, well-drained soil. Water deeply in dry weather. Fertilize in spring every 1–2 years.

KALANCHOE The best-known *kalanchoes* are succulents to 2′ with scarlet flower clusters in winter. They are popular house plants which also grow outdoors in warmest climates.

For how to grow, see cactus, desert. Add a little extra humus to the soil. Give an indoor winter temperature of 60°. For best flowering, grow new plants from seeds sown every March in sterilized soil. Grow the young plants first in a north window; then in summer in an east window (or its equivalent outdoors); then in a south window (or full sun) in fall and winter. Since *kalanchoe* is a short-day plant, avoid exposure to electric light in the evening during late fall and winter.

KALE *Brassica oleracea acephala.* A type of cabbage to 30″ with long, upright, curly, ruffled leaves used as "greens." Grows almost everywhere in cool weather, but is usually grown in the south.

Sow seeds in July or August in well-drained, limed, fertilized soil. The rows should be 2′ apart and plants thinned to 12″–15″ apart. Keep weeded and watered. If soil is not very fertile, sidedress once with fertilizer rich in nitrogen. Plants continue to grow when outer leaves are picked off for cooking. In the south, plants can be left outdoors through the winter.

KANGAROO VINE *Cissus antarctica.* Evergreen vine to 10′, a favorite of house-plant growers. Also occasionally grown outdoors in subtropics.

Indoors. Propagate by stem cuttings and plant in pots of general-purpose soil. Grow in a north window. Water when soil feels dry. Fertilize monthly while vine is making growth. Train on strings or a light trellis. Prune stems that grow too long.

Outdoors. Grow like evergreen grape (which see), but give more shade.

KATSURA TREE *Cercidiphyllum japonicum.* Deciduous tree to 60′ with branches that reach upward and fine fall foliage. Grows in mild climates.

Plant in spring in an open location where tree can grow upward and outward. The plant needs better-than-average soil. Keep reasonably moist. Fertilize every 3–4 years.

KENILWORTH IVY *Cymbalaria muralis,* coliseum ivy, Aaron's beard, climbing sailor. Climbing or creeping ivy about 3′ long with little pink, white or blue flowers. Mainly a house plant.

Pot in general-purpose potting soil. Grow in an east, west or south window. Keep moist. Spray foliage with water frequently. Fertilize every two months while making growth. This is a good plant for a hanging basket and can be moved outdoors in summer into the partial shade of the terrace.

KENTUCKY COFFEE TREE *Gymnocladus dioica.* Deciduous tree to 90′ with large seed pods. Grows mainly in mild climates east of the Mississippi.

Plant in early spring or fall in sun. Grow in average soil which holds moisture. Water and fertilize annually for the first 2–3 years. After that, tree needs little attention.

KERRIA JAPONICA Japanese rose, globeflower. Deciduous shrub to 8′ with bright yellow flowers in spring. Grows in mild and warm climates.

Propagate by stem cuttings or root division. Plant in fall or spring in average soil in a sheltered spot that is sunny or partially shaded. Water in dry weather. Fertilize lightly every 1–2 years. Prune out old stems after flowering. Keep suckers cut out. Tips of young branches may be killed in severe winters, but this does not harm the plant.

KLEINIA Different species known as candle plant, inchworm plant, blue chalk sticks. Succulent shrubs or trailers to 10′ with thistle-like, yellow or white flowers. Grows in warmest climates; also indoors.

For how to grow, see cactus, desert. Provide very well-drained soil, and water very sparingly when plants are resting.

KOCHIA Burning bush, summer cypress, Mexican firebush. Erect 30″ annual resembling an evergreen, with red foliage in the fall. Grows almost everywhere.

Soak seeds in water for 24 hours, then sow indoors about eight weeks before last frost. Transplant into average soil in full sun. Fertilize when plants are established. Water regularly.

KOHLERIA Incorrectly identified as *isoloma*. Plant growing from a rhizome to about 2′, with tubular flowers in several colors. A house plant.

Grow like gloxinia (which see). Propagate by young tip cuttings or by division of the rhizomes.

KOHLRABI *Brassica caulorapa*, turnip cabbage. A 1′ member of the cabbage family that produces a creamy or purplish, turnip-like bulb just above ground. Grows best in cool climates.

Grow like cabbage (which see). Plants, however, should be spaced 8″ apart in rows 18″ apart. Harvest bulbs when they are about 2″ across and still growing.

KUDZU *Pueraria thunbergiana*, Jack-and-the-beanstalk. Very fast-growing perennial vine to 70′. It is often used as a groundcover. Grows best in the south.

This vine easily becomes a dreadful pest, because it spreads like wildfire. However, it is useful in holding soil on steep banks and for quick screening. Propagate by root division. Plant in average soil without much moisture or fertilizer. Grow in sun or shade. Pinch out stem ends during spring and summer and cut back severely in the fall.

KUMQUAT *Fortunella*. Evergreen trees or shrubs to 25′, bearing small, orange-like, edible fruits. Grows in warmest climates.

Grow like orange (which see).

LADY FERN *Athyrium filix-femina*. Lovely fern with bright green, feathery fronds 30″ long. Grows mainly in the east.

For how to grow, see fern. Lady fern grows in sun or shade. It prefers moist soil but tolerates dry.

LADY PALM *Rhapis excelsa*. Palm trees to 15′ with slender stems and open, fan-shaped fronds reminiscent of *clivia* foliage. Grows in subtropics; also indoors.

For how to grow, see palm. Plant the lady palm in partial shade, although it does well in sun. If grown as a hedge, keep new suckers from spreading beyond the hedge row, and cut out old shoots when they grow too tall; but do not shear.

LADYSLIPPER *Cypripedium*. One species called moccasin flower. Perennial wildflowers to 2′ with pink, yellow or white, pouch-shaped flowers. Grows mainly in the north.

Dig up from the wild with a good ball of soil, or divide roots into sections containing at least one bud. Plant in early spring or fall in light shade. *C. acaule* requires a very acid, well-drained, dry, humusy soil. *C. candidum* needs a moist, sweet, humusy soil. All others need a moist, humusy, neutral or slightly acid soil. In no cases are the plants particularly easy to grow, and they often do not last very long.

LAMB'S EARS *Stachys lanata*, woolly woundwort. Perennial to 18″ with white, woolly hair on foliage and purple flower spikes. Grows in temperate climates.

Divide in spring or fall and plant in sun in average soil. Water in dry weather. Fertilize in the spring.

LANTANA Evergreen and deciduous shrubs to 4′ with flat-topped flower clusters of various colors much of the year. Grows indoors; also outdoors in warmest climates.

Indoors. Propagate by stem cuttings rooted in sand and peatmoss in the spring. Pot up in general-purpose potting soil and grow outdoors in a sunny spot through the summer. Keep buds pinched off. Bring indoors in the fall and grow in a south window at about 65°. Water regularly and fertilize lightly every other month. Don't try to carry over plants from year to year; start new cuttings each spring instead.

Outdoors. Plant in spring or fall in a sunny location in average soil. Water only in dry weather and then not too much. Fertilize once in early spring. Prune moderately in early spring.

In cold climates where *lantana* cannot survive, the plants can nevertheless be grown in the sunny flower bed during the summer. Bring inside in the fall as a house plant.

LAPEYROUSIA Sometimes identified as *anomatheca*. Plants growing from corms to 20″ with small red, white or blue flowers in summer. Grows in temperate climates.

Grow like gladiolus (which see). Plant corms 3″–4″ deep.

LARCH *Larix*, tamarack, hackmatack. Coniferous trees to 100′ with clusters of needles which are shed in the fall. Grows in the northeast, northwest and Great Lakes region.

Plant in spring or early fall in an open location. Grow in average soil which is well drained but holds moisture. Water deeply in dry weather. Fertilize every 3–5 years. Spray in spring with malathion to control sawflies, which defoliate the trees. Cut out oozing cankers that may develop in the trunks, and coat the wounds with tree paint.

LARKSPUR *Delphinium.* Erect annual to 4′ with handsome blue, pink, red or white flower spikes in spring. Grows best in cool climates.

Sow seeds where plants are to grow as soon as the soil can be worked in the spring, or sow in the same location in late fall. Larkspur needs sun and average well-drained soil. Fertilize when plants are thinned to 9″. Cultivate lightly and keep watered. A second light feeding may be helpful just before plants start to bloom.

LAUREL CHERRY *Prunus caroliniana*, cherry laurel. Sometimes identified as *Laurocerasus caroliniana.* Evergreen shrub or tree to 25′ with glossy leaves and white flowers in spikes in late winter. Grows in warmest climates.

Plant in spring or fall in sun in average soil. Water in dry weather. Fertilize every 2–3 years in winter. Prune after flowering.

LAVENDER *Lavandula.* Perennials to 6′ with fragrant leaves and lavender flowers. Grows almost everywhere.

Propagate by cuttings in the spring. Grow in a sunny spot in light, well-drained, average soil. Water regularly. Fertilize in spring. Cut back after flowering. Dry leaves and flowers in a shady spot.

LAVATERA Tree mallow. Bushy annual to 5′ with hollyhock-like flowers. Grows everywhere.

Sow seeds outdoors after last frost in full sun in average soil. Water in dry weather. Fertilize once or twice. Pick off dead blooms.

LEATHERLEAF *Chamaedaphne calyculata.* Evergreen shrub to 5′ with white flower clusters in early spring. Grows in mild and cold climates.

Propagate by layering. Plant in spring or fall in sun. The soil should contain considerable sand and humus. Give plenty of water. Fertilize every 2–3 years.

LEEK *Allium porrum.* Two-foot member of the onion family that does not form a bulb. The 3′ slender stem is of very mild flavor. Grows wherever warm weather continues for at least 130 days.

Leeks grow in average, well-drained, humusy soil that is limed and enriched with general-purpose fertilizer. Sow seeds in early spring in rows 2′ apart. Thin plants to 3″–4″. Sidedress 2–3 times with fertilizer. Cultivate and weed carefully. Water well. Give full sun. As the plants grow large, hill up soil 6″ around them to blanch the stalks.

LEMAIREOCEREUS Organ pipe cactus. Different species also known as Arizona organ pipe, blue miter. Cacti to 20′ with stout, spiny stems rising from the ground. Grows in hot, dry climates.

For how to grow, see cactus, desert.

LEMON *Citrus limon.* Evergreen trees to 25′ with sour, yellow fruits. Grows mainly in warm parts of California. Small varieties that grow well in the house are Meyer and ponderosa.

Grow like orange (which see).

LEMON BALM *Melissa officinalis,* balm. Perennial herb to 3′ grown for its leaves, which are used in flavoring. Grows in mild climates.

Divide plants in spring, or sow seeds in summer and move the young plants into their permanent location the next spring. Grow in a sunny, sheltered spot in average soil. Water in dry weather. Fertilize in spring.

LEMON VERBENA *Lippia citriodora.* Shrub to 10′ with fragrant foliage and white flowers in summer. Grows outdoors in warmest climates.

Plant in spring or fall in sun in average soil. Water in dry spells. Fertilize every 1–2 years in the spring. Prune hard in the fall. In cold areas, bring plants into a bright room in winter. Water sparingly.

LEOPARD PLANT *Ligularia tussilaginea aureo-maculata.* Perennial to 2′ with big, spotted leaves and yellow summer flowers. Grows indoors; also outdoors in warm climates.

Indoors. Propagate by division in early spring. Pot in general-purpose potting soil. Keep moist. Grow in a cool east or west window. Fertilize in spring and every couple of months through the summer.

Outdoors. Plant in spring or fall in average, well-drained soil in partial shade. Water in dry weather. Fertilize in early spring.

LEOPARD'S BANE *Doronicum.* Perennials to 30″ with yellow, daisy-like flowers in spring. Grows in temperate climates.

Divide plants every other year after they have bloomed. Grow in sun or light shade in average soil improved with humus. Fertilize in early spring Water regularly.

LESPEDEZA Bush clover. The best of the lespedezas are the shrubs, to 9′, which bloom rather profusely in late summer, and *L. striata,* Japan clover, which is used as a cover crop. The shrubs grow in mild and warm climates; Japan clover in the south.

Shrubs. Propagate by stem cuttings. Plant in spring or fall in a sunny spot in sandy soil. Water in very dry weather. Fertilize only occasionally.

Japan clover. Sow seeds in the spring and plow under in the fall.

LETTUCE *Lactuca sativa.* Loose, or leaf, types of this popular vegetable are easier to grow than head lettuces. Grows almost everywhere but does not tolerate heat.

Grow in average, neutral soil in full sun. For earliest crop, sow seeds indoors 8–10 weeks before last frost. Grow in a cool, sunny window. Or sow seeds outdoors as soon as soil can be worked. Make several sowings at two-week intervals for a continuing supply.

Lettuce should be grown in rows 15″ apart. Thin plants to 10″ and sidedress with general-purpose fertilizer at that time. Keep well watered.

LEUCOPHYLLUM TEXANUM Barometer bush, Texas ranger. Evergreen shrub to 12′ with purple summer flowers. Grows in warmest, dry climates.

Plant in spring or fall in full sun in average, well-drained soil that is neutral or alkaline. Prune in winter. Needs little attention otherwise.

LEUCOTHOE Different species called fetter bush; pepper bush, sweetbells or white osier. Shrubs, usually evergreen, to 12′, with handsome leaves and drooping sprays of lily-of-the-valley-like, white flowers in spring. Grows in mild and warm climates.

Propagate by stem cuttings in spring. Plant in spring or fall in partial shade in moist, acid soil containing considerable peatmoss and sand. Water in dry spells. Fertilize annually in early spring with a 10-8-8 fertilizer. Pick off flower heads and prune as necessary at that time.

LIATRIS Gayfeather, button snakeroot, blazing star, colicroot, rattlesnake master, prairie pine, devil's bit. Perennials to 6′ with reddish-purple flower spikes in summer. Grows almost everywhere.

Propagate by division in spring and plant in sun or partial shade in average, well-drained soil. Water regularly. Fertilize in spring.

LILAC *Syringa.* Deciduous shrubs to 30′ with gorgeous, fragrant flowers in all colors except yellow and orange. Grows best in mild and cold climates, though the Persian lilac does pretty well in warm climates.

If lilacs are purchased, ask whether they are own-root plants or grafted on to the roots of other plants. Lilacs are propagated by the suckers of "own-root" plants or by greenwood stem cuttings taken from any plant in May or June.

Plant lilacs in very early spring or fall in a sunny location. The soil should be deeply dug, well-drained, limed if acid, and enriched with dried manure or bone meal. Set own-root plants in the ground at the depth they previously grew, but bury the graft of grafted plants. Keep well watered for the first six months; thereafter, water only in dry spells. Apply bone meal in early spring for the first two years; then every other year. Apply lime every other year if needed.

When cutting flowers, take long, leafy stems with them. Remove faded flower clusters. Keep suckers cut out. Once every 2–3 years, in winter, remove branches that do not have vigorous flower buds.

Dust with powdered sulfur if mildew is a problem. Apply a dormant oil spray just before the buds break in the spring to control scale. Spray with malathion in June and July if borers appear.

LILY *Lilium.* Favorite bulbous plants to 8′ with magnificent flowers in many colors from spring through September, depending on the type. Grows almost everywhere, but is not too happy in warmest climates. Also, to some extent, a house plant.

Note. Lily species are beautiful and difficult. Grow hybrid lilies only.

Outdoors. Plant the bulbs right after you buy them in the fall in sun in a place with good air circulation (not close up against a wall, for example). The soil can be of average quality, but must be deeply dug and mixed with a great deal of peatmoss or humus from oak leaves or pine needles. (Lilies prefer soil that is somewhat acid.) Space bulbs 12″–18″ apart and cover with 4″ of soil (Madonna lilies, however, should be only 1″ deep). If you've ever had rodents in the garden, enclose the bulbs in ½″ wire mesh. Scratch a little fertilizer into the soil and cover the soil with a mulch of peatmoss. Maintain this mulch the year round: lily roots should be kept cool in summer and protected in winter.

In early spring and again about two months later fertilize the bulbs lightly. Water in dry weather so that the soil is evenly moist at all times. If you cut flowers for arrangements, leave as much foliage on the plant as possible. Pick off dead flowers. Allow foliage to die down naturally.

Divide lily clumps that become crowded after flowering. Spray with spray used for roses if insects are a nuisance.

Indoors. Lilies are not easily grown in the house, but you can try. Use the Croft variety of *L. longiflorum.* Plant bulbs 1″–2″ deep in separate pots in general-purpose potting soil. Water and keep in a dark place at 45° until growth starts. Then gradually move plants into a south window at 60°. Keep soil evenly moist. Fertilize every three weeks.

To save lilies that come from the florist, keep them watered and in a cool, sunny window. Then plant outdoors after all danger of frost is over.

LILY OF THE VALLEY *Convallaria.* Favorite spring flower which grows to 9″ from rootstocks known as pips. Grows in temperate climates; also indoors.

Outdoors. Plant the pips 1½″ deep in spring or fall in a shady spot. The soil should be well-drained, rich, humusy and moist. Fertilize in early spring. Dig up and replant every 2–3 years; otherwise the plants may crowd themselves out of existence.

Indoors. Lily of the valley can be grown successfully indoors only if you start with large pips that have been prepared for forcing by growers who stored them at low temperature. Plant these forcing pips in sphagnum moss so the tips just show. Moisten the sphagnum thoroughly and place the container in a closet for a week. Keep watered. Then gradually expose to more and more light. About two weeks after planting, place container in south window. Flowering should start about a week later. It won't continue long.

LILY TURF *Ophiopogon japonicus,* mondo grass. Evergreen plant to 1′ with grass-like leaves and small purple flowers. Used as a groundcover in warm climates. See also *liriope.*

Divide in spring and plant in average, moist soil. Space small clumps about 3″ apart; larger ones up to 6″. Lily turf grows in sun but is especially useful in shade. Water in very dry weather. An occasional application of fertilizer will stimulate growth but is rarely needed.

LIMA BEAN *Phaseolus limensis,* butter bean. Slow-maturing bush or pole beans (to 10′) grown for their large, succulent seeds. Grows almost everywhere.

Grow like bean—green, snap or string (which see). However, seeds of bush varieties should be planted 3″–4″ apart in rows 3′ apart. Because of slow maturity, succession sowings are not very practical in the north.

LIME *Citrus Auranti folia.* Evergreen trees to 25′ with small, green fruits. Grows mainly in southern Florida.

Grow like orange (which see). Space West Indian limes 20′ apart; Persian limes, 25′.

LIMONIUM Sea lavender, sea pink. Incorrectly identified as *statice.* Annuals, biennials and perennials to 3′ with small flower clusters in several colors. Used in dried arrangements. Grows in warmest climates.

Sow annual seeds in the spring where plants are to grow. Propagate perennials by division in the spring. Grow in average soil, preferably one that is quite sandy. Give full sun. Water in dry weather. Fertilize in the spring.

LINARIA Toadflax. Annuals and perennials to 4′ with delicate, snapdragon-like flowers in yellow, blue, purples, white. Grows almost everywhere.

Grow in full sun in average soil which is kept reasonably moist and is fertilized a couple of times during the growing season. Annual *linaria* is started from seeds sown indoors 6–8 weeks before last frost, or outdoors when weather is warm. Perennials are propagated by root division in spring or fall.

LINDEN *Tilia*, basswood. Handsome, deciduous trees to 120′ with heart-shaped leaves and fragrant, yellowish-white flowers. Grows in moist, temperate regions.

Plant in spring or fall in sunny locations. The trees grow in average soil, but the richer and more moisture-retentive the soil, the better the results. Water deeply in dry spells. Fertilize every 3–5 years.

LIRIOPE Different species called blue lily turf or creeping lily turf. Grass-like, evergreen groundcover to 18″ with terminal clusters of blue or pink flowers. Grows in warm climates. See also lily turf.

Divide in spring and plant in shade in average, moist soil. Fertilize in spring if plants need a boost.

LITTLE PICKLES *Othonna crassifolia*. Succulent with short, trailing or creeping stems, tiny, fleshy leaves and yellow, daisy-like flowers most of the year. A house plant, but also grows outdoors in warmest climates.

For how to grow, see cactus, desert. Water very carefully, because too much moisture encourages rotting. Needs full sun to bloom.

LIVING ROCKS Desert cacti of various genera, usually spineless. Plants closely resemble rocks a few inches across. Grows indoors; also outdoors in hot, dry climates.

For how to grow, see cactus, desert.

LIZARD'S TAIL *Saururus cernuus*, American swamp lily. Perennial to 5′ with large, heart-shaped leaves and slender, white flower spikes. Grows almost everywhere in marshes.

Divide roots in spring and plant in half shade in average soil. Keep wet or even flood the site permanently with up to 2″ of water.

LOBELIA One species known as cardinal flower or Indian pink. The perennial species grow up to 5′, and have clusters of blue or red flowers. The only commonly grown annual is an 8″ edging plant usually with blue flowers. Grows almost everywhere.

Perennial. Propagate by division in the fall, or sow seeds in midspring. Grow in partial shade in average soil. Keep moist. Fertilize in early spring and again about June.

Annual. Sow seeds indoors 8–10 weeks before last frost. Transplant into average soil in partial shade. Fertilize when young plants are established. Keep watered.

LOCUST *Robinia*. Open, deciduous trees to 80′ with pendulous clusters of small pink, white or purple flowers in spring. Grows mainly in the east. See also rose acacia.

Plant in early spring or fall in full sun in average, well-drained soil. Water in drought. Fertilize every 3–5 years. Prune out dead and broken wood and rip out the suckers, which crop up everywhere.

LOGANBERRY Bramble fruit with red fruits like blackberries. Grows in warm climates.

Grow like blackberry (which see). However, plants should be spaced about 8′ apart in the row, and trained to horizontal wires.

LONDON PLANE TREE *Platanus acerifolia.* Deciduous tree to 140′ with big leaves and bark that peels off in plates. Commonly grown in the east and other mild climates.

Plant in spring in a sunny location. Ideally, the soil should be moist and of above-average quality; however, the tree does very well in ordinary soil and with little care. (This is a good street tree.) Water in long dry spells. Fertilize every 3–5 years.

LOOSESTRIFE This name is applied to *lysimachia, lythrum, decodon* and *steironema.* Perennials of varying habit, to 8′, with flowers in various colors. Grows almost everywhere.

Propagate by divisions in the spring and plant in light shade or sun. The plant's most important requirement is moist, fairly rich, humusy soil (as along a stream bank). Fertilize in spring.

LOQUAT *Eriobotrya japonica.* Evergreen tree to 20′ with fragrant, white flowers and edible, acid, plum-shaped, yellow fruits. Grows in warmest climates.

Plant in early spring or fall in sun. Grow in average, well-drained soil. Water in dry weather. Fertilize annually before flowering starts. Mixing some humus or dried manure into the soil every year is advisable though not essential. The size of the fruits is increased somewhat if some of them are thinned out. No special pruning is required.

LOROPETALUM CHINENSE Evergreen shrub to 10′ with blue-green foliage and small, white, spring flowers. Grows in warm climates.

Plant in spring or fall in partial shade in average soil. Keep well watered. Fertilize every 2–3 years in late winter. Prune after flowering. The plant looks its best when kept low.

LOTUS *Nelumbo.* Aquatic plants with large, bowl-shaped leaves, huge flowers borne high above the water, and funnel-shaped pods used in flower arrangements. Grows almost everywhere.

Grow like hardy water lilies (see water lily). But note the following differences: Lotus can be grown in very shallow water. A little additional manure improves growth. Repot and divide roots every two years. If the plant is likely to be frozen in ice in the winter, store the container indoors at a temperature of no more than 40°. Keep soil moist during storage and cover the container with wire mesh to keep out mice.

LOVAGE *Levisticum officinale.* Perennial herb to 6′, grown for its celery-flavored leaves. Grows in warm and mild climates.

Divide in spring and plant in rich soil in sun. Water well. Fertilize in spring.

LUPINE *Lupinus.* Beautiful plants to 5′ with symmetrical flower spikes in various colors. The best of the annuals, *L. hartwegi,* grows widely throughout the country but prefers cool climates. Two favorite annual wildings are the Texas bluebonnet and Quaker bonnets found in the east. The best of the perennials, the Russell hybrids, grow satisfactorily only in cool, moist climates such as New England and the northwest.

Perennial. Sow seeds in a seedbed in late summer and move plants in early spring to their permanent location. Lupines do not like to be disturbed once well established. Grow in a well-drained, light, slightly humusy, neutral or slightly acid soil in sun or partial shade. Avoid very hot locations. Do not let soil dry out for long periods. Fertilize in early spring and once or twice again with plant food rich in phosphorus. Pick off fading flowers. When flowers go, cut off stalks but leave as much foliage as possible. Mulch in winter.

Annual. Sow seeds where plants are to grow after soil is warm. Handle as above.

LYCASTE SKINNERI Semi-deciduous, epiphytic orchid to 15″ with large, waxy, spring flowers which are white and often tinged with rose or yellow. A house plant.

For how to grow, see orchid. Move plant outdoors in summer into light shade.

LYCHNIS Campion. Different species also known as Maltese cross, Jerusalem cross, scarlet lightning; rose of heaven; mullein pink, dusty miller or rose campion; cuckoo flower or ragged robin; flower of Jove; German catchfly. An annual variety is often identified as *viscaria*; a perennial as *agrostemma.* Annuals, biennials and perennials to 30″ with mainly red flowers which grow best in cool climates.

Almost all species will flower the first year if seeds are sown indoors about eight weeks before last frost. Transplant into average, well-drained soil in full sun. Fertilize when well established. Water in dry weather. Divide perennials in the spring when clumps become large. Many species self-seed wildly.

LYCORIS Magic lily, hurricane lily, spider lily, hardy amaryllis. Two-foot bulbous plant with circular clusters of flowers in many colors at a time different from that when foliage appears. Mainly warm-climate plants, but *L. squamigera* grows in the north.

Plant bulbs 4″–5″ deep and 8″ apart in good, well-drained soil in the fall. The flowers do best in light shade but also bloom in sun. Fertilize when they are sending up leaves. Water in long dry spells.

LYSIMACHIA NUMMULARIA Moneywort, creeping Charlie, creeping Jenny. Prostrate perennial only a few inches tall with yellow flowers. A groundcover in temperate climates.

Divide in spring or fall and plant in partial shade or sun in average soil. Water in dry weather. Fertilize in spring.

MAACKIA AMURENSE Amur yellow wood. Deciduous tree to 40′ with white flowers in erect clusters in summer. Grows in all but extreme climates.

Plant in spring or fall in a sunny location that is somewhat protected from wind. Grow in average soil. Water deeply in dry weather. Fertilize every 3–4 years in spring.

MACADAMIA TERNIFOLIA Queensland nut. Vase-shaped, evergreen tree to 25′ with holly-like leaves and edible nuts after tree is a number of years old. Subtropical.

Plant in spring or fall in sun. The soil must be well-drained, rich and very humusy. Add lime if soil is acid. Water regularly in all dry spells. Fertilize in early winter. Prune after bearing in summer.

MACLEAYA Plume poppy, tree celandine. Sometimes identified as *bocconia.* Perennial to 8′ with handsome leaves and plume-like panicles of white flowers in summer. Grows in mild and warm climates.

Divide every three years in spring and plant in sun in good soil that is fertilized in early spring and again in June. Keep watered. Protect against strong wind.

MADAGASCAR JASMINE *Stephanotis floribunda.* Evergreen vine to 15′ with clusters of fragrant, waxy, white flowers in summer. Grows outdoors in warmest climates; also indoors.

Propagate by semi-ripe stem cuttings in March. Plant in well-drained, general-purpose potting soil in a pot or in the garden, and grow in partial shade or an east or west window. Keep moist and feed while making growth in spring and summer at three-week intervals. Stop fertilizing and gradually reduce water in fall; and in winter, water only enough to keep soil from drying out completely. During this time, a vine grown indoors should be kept in a cool (down to 60°) room. Eliminate weak wood and cut back strong stems as necessary in early spring. Train on a trellis.

MADEIRA VINE *Boussingaultia baselloides*, mignonette vine. Tuberous-rooted, evergreen vine to 30′ with lush foliage and fragrant, white flower spikes in late summer. A perennial in warm climates; handled as an annual further north.

Propagate by dividing the tuber or by the little tubers at the juncture of leaves and stems. Grow in full sun in average, well-drained soil. Don't over-water. Fertilize when growth starts. Grow on a trellis that tendrils can wrap around. In warm climates, where tubers can be left in the ground, prune the vine in early spring. Elsewhere, after foliage has withered, cut the plant back to the ground, lift the tuber and store like a dahlia tuber (see dahlia).

MAGNOLIA Different species known as sweet bay, swamp bay, white bay, beaver tree, swamp laurel; cucumber tree; umbrella tree; laurel sabino; bull bay; yulan. Handsome deciduous and evergreen trees to 100′, with showy white, purple, pink or yellow flowers in spring and summer. Deciduous types grow in all but the coldest climates; evergreens grow in warm climates.

Deciduous. Plant small specimens only in early spring in sun or partial shade. The soil should be better than average, containing sand and humus. Water in dry weather. Fertilize every 1–2 years. Avoid disturbing established plants. Prune after flowering only to remove dead wood. New plants are propagated by planting seeds outdoors as soon as they ripen.

Evergreen. Handle as above but plant where trees will receive summer sun but will be protected against winter sun and winds. To propagate, harvest ripe seeds and store in dry sand until about Februrary. Then moisten sand for two weeks. Then remove seeds and wash, and plant in pots of general-purpose potting soil with a little extra humus. Move to a larger pot when young plant is well developed. Then set out in the garden in early spring.

MAGNOLIA VINE *Schisandra propinqua.* Deciduous vine to 20′ with orange flowers and clusters of red berries in fall. Grows in southeast.

Propagate by layering. Plant in good soil in partial shade. Protect from wind. Keep watered. Fertilize and prune in spring. Provide sturdy trellis. To have berries you must plant both male and female specimens.

MAIDENHAIR FERN *Adiantum pedatum* grows outdoors in many parts of the country. *A. cuneatum* is a house plant. Favorite, lovely ferns about 2′ tall with delicate, more or less fan-shaped leaf segments and black stems.

For how to grow, see fern. Outdoors, the maidenhair very definitely needs shade and a cool, light, humusy, moist but well-drained soil. Indoors, the plant is somewhat difficult because it needs high humidity and lower-than-normal house temperature. Grow in a north window.

MALPIGHIA One species called Barbados cherry is an evergreen shrub to 10′ with glossy, red, cherry-size, edible fruits. Another species called miniature holly is a 3′ shrub with leaves like holly, pink flowers and red fruits. Both grow outdoors in subtropics. The second is also a house plant.

Outdoors. Plant in sun in average, well-drained soil in the fall. (Miniature holly also grows in partial shade.) The soil must first be fumigated, because nematodes attack the roots. Water in dry weather. Fertilize in late winter every year. Prune sparingly—especially Barbados cherry—because fruits are borne on old wood.

Indoors. Pot miniature holly in general-purpose potting soil. Keep in an east or west window down to 55° at night. Water thoroughly when soil feels dry. Fertilize every two months.

MANDEVILLA SUAVEOLENS Chilean jasmine. Sparse deciduous vine to 15′ with profuse, white, deliciously fragrant flowers in summer. Grows in warm climates.

Propagate by stem cuttings or seeds. Plant in rich, sandy loam with humus and fertilizer added. Give full sun except in very hot areas, where partial shade is better. Keep constantly watered except when plant is dormant. Fertilize established plants three times—in spring, summer and fall. Train to a sturdy support. Pick off seed pods when they form. Prune lightly after flowering to control and shape growth. Spray with malathion if plant is attacked by red spider.

MANGO *Mangifera indica.* Evergreen tree to 90′ with long leaves, pink flower clusters and large, juicy, red or yellowish fruits. Grows in subtropics.

Plant in spring in a sunny location in average, well-drained soil. Space trees 40′ apart. Water heavily every two weeks after the fruit has set in late winter and early spring. Apply 1 lb. of ammonium sulfate per year of age of the tree in winter when flower clusters are forming but before they open. Then, a month later, apply a general-purpose fertilizer recommended for your particular soil by the county agent.

Spray with copper fungicide when bloom starts to control anthracnose. Repeat application when fruit starts to set. To protect trees up to four years old against cold, wrap trunks up to the bottom branches with 3″ or more of straw.

MANZANITA *Arctostaphylos.* Evergreen shrubs to 20′ with bell-shaped, white or pink flowers, red berries and purplish bark. Grows on West Coast.

Plant in early spring or fall in sun or light shade in average soil that is well drained. Don't water in summer, but at other times, water in dry spells.

MAPLE *Acer.* One species called box elder; another, moosewood. Deciduous trees to 120′ with handsome autumn foliage. One species or another is to be found almost everywhere.

Plant in early spring or after leaves drop in the fall. Grow all maples in sun except the Japanese maple, which tolerates light shade. The soil need be of only average quality but must be well drained. Water in the spring if the winter has been very dry; also water in other long dry spells. Fertilize every 3–5 years.

MARANTA Different species called arrowroot, prayer plant. Plant to 6′ with variegated, nicely colored leaves. Grows indoors.

Divide in early spring. Pot in equal parts of loam, humus and sand. Grow in a warm east or west window. Keep soil moist but not soggy at all times except during the late fall and early winter, when supply can be reduced substantially. Spray foliage with water frequently. Fertilize every 4–6 weeks from spring through summer.

MARGUERITE *Chrysanthemum frutescens.* Boston daisy, Paris daisy. Perennial to 3′ with yellow or white daisies. A house plant. Also grows outdoors in warm and cool climates.

Indoors. Buy plants in summer and grow outdoors in pots in general-purpose potting soil. Bring indoors before frost and grow in a cool south window. Water well. Fertilize every two weeks. Don't try to carry over old plants.

Outdoors. Plant purchased plants in sun in average soil. Pinch stem end to make plants bushy. Pick off dead flowers. Replace plants every 2–3 years, but best treated as an annual.

MARIGOLD *Tagetes.* Ever popular annual to 3′ that does well almost everywhere. Some of the French marigolds also grow indoors.

Outdoors. For earliest flowers, sow seeds indoors 4–6 weeks before last frost and shift outdoors when weather has settled. Or sow seeds where plants are to grow after danger of frost is past. In either case, grow in sun in average, well-drained soil. Water in very dry weather. Fertilize when plants are established in the garden and once again if the soil is not overly good. When plants are young, pinch growing tips often to promote bushy growth. Pick off dead flowers.

Indoors. Sow seeds in pots outdoors in late summer. Bring inside before frost and keep in a cool, sunny window. Water when soil feels dry.

MARIPOSA LILY *Calochortus,* globe tulip, butterfly tulip. Different species also called sego lily, cat's ear and fairy lantern. Plants growing from corms to 3′ with flowers in various colors and shaped somewhat like tulips. Grows best in California, but also east to Nebraska.

Plant corms to twice their depth and about 4″ apart in the fall in partial shade. The soil should be well-drained loam mixed with considerable sand and humus. Water while making growth in late winter and spring, but do not water in summer.

MARSH MARIGOLD *Caltha palustris*, cowslip. Perennial to 2′ with more or less round leaves and large, buttercup-like flowers in very early spring. Grows east of the Rockies in cool, wet places.

Propagate by division of the roots after flowering. Plant in light shade in average soil with extra humus. Keep well watered at all times. Fertilize a little in the spring.

MATRIMONY VINE *Lycium*. Prickly, deciduous vine to 8′ with orange-red berries in fall and winter Grows in all but coldest climates.

Propagate by layering. This plant can become a nuisance, so plant it in an out-of-the-way place in spring or fall. Grow in sun or partial shade in average, well-drained soil. Water only in driest weather. Keep suckers cut out. Thin rest of plant in the spring. If used as a climber, tie branches to a sturdy support. You may also use the plant as a cover for unsightly banks, rock piles, etc.

MAURANDIA Slender-stemmed vine to 10′ with nice foliage and a profusion of purple or rose flowers in summer. A perennial, but if treated as an annual it can be grown almost everywhere.

Sow seeds indoors eight weeks before last frost. Shift into individual pots filled with general-purpose potting soil to which a little humus has been added. Fertilize at that time and monthly thereafter through the summer. Since *maurandia* blooms best when confined in pots, move pots outdoors into full sun or partial shade when weather is warm. Protect from winds. Keep soil evenly moist. Spray with malathion to control red spider and aphids. Train stems on string, wire or a light wood lattice.

MAYAPPLE *Podophyllum peltatum*, mandrake, Indian apple. Creeping perennial wildflower to 18″ with large, umbrella-like leaves hiding white flowers which are followed by edible yellow fruits. Grows mainly east of the Mississippi.

Divide roots and plant in early spring or fall in partial shade in average, well-drained soil containing extra humus. Water in dry weather. Fertilize in early spring.

MAYTEN TREE *Maytensis boaria*. Evergreen tree to 40′ looking like a weeping willow. Grows in warm climates.

Plant in spring or fall in sun in average soil. Water in dry spells. Fertilize in late winter every 3–4 years.

MEADOW BEAUTY *Rhexia*. One species called deer grass. Perennial bog plant to 2′ with oval leaves and purple summer flowers. Grows east of the Mississippi.

Propagate by stem cuttings in late spring. Plant in the fall in light shade in very acid, light, humusy, always moist soil.

MEADOW RUE *Thalictrum.* Perennials to 8′ with delicate foliage and large clusters of summer flowers in various colors. Grows mainly in temperate climates.

Divide plants every 3–4 years in spring and plant in sun in average soil containing a little extra humus. Water regularly. Fertilize in spring.

MERTENSIA Virginia bluebells, Virginia cowslip, lungwort. Perennials to 2′ with clusters of bright blue flowers in early spring. Most species grow in cool climates, but *M. virginica* also does well in warm areas.

Sow seeds as soon as possible after they are harvested. Transplant in the fall into rich soil with considerable humus. Grow in partial shade. Keep moist. Fertilize in early spring. Growth dies down early so take care in cultivating around plants. Better mark them and don't move them once they are established.

MESEMBRYANTHEMUM Different species known as ice plant, hottentot fig, sea fig, tongue leaf, elk's horns, Shriner's plant, desert rose, tiger jaws, lobster claws, victory plant, African living rock, living stones, split rock, silverskin, cone plant, karroo rose, shark's head, jewel plant, baby toes. Succulents to 3′ of many different genera and several different forms, with brilliant daisy-like flowers in many colors. Grows in hot, dry climates; also indoors.

Grow the trailing and shrubby mesembryanthemums like cacti (see cactus, desert). The other types require a soil mixture of one part loam, one part humus, two parts sand and one part gravel. Be careful not to plant too deeply lest rot set in. Water when soil feels dry while plants are making growth, but water very sparingly during rest periods. Grow in full sun. Fertilize lightly at the start of growing periods.

MEXICAN BAMBOO *Polygonum cuspidatum.* Eight-foot perennial with green, bamboo-like stems and countless tiny pink flowers in late fall. Grows in all but coldest climates.

The common white-flowered variety is a dreadful pest, but the pink-flowered variety is delightful and does not spread. Propagate it by division of the roots in fall or spring. Plant in average soil. It needs no care but responds well to watering in dry weather and a little plant food every 1–2 years. Cut stalks to the ground in the fall when flowers are deepest pink, and keep them in a dry place outdoors until flowers and leaves are dry. Then pick off leaves and use flowers indoors in arrangements. They last for years.

MEXICAN FLAME VINE *Senecio confusus.* Thin-stemmed, evergreen vine to 25′ with profuse orange-red flowers. Grows outdoors in warm climates; also indoors.

Grow like German ivy (which see).

MEXICAN FOXGLOVE *Allophyton mexicanum.* Also identified as *Tetranema mexicanum.* Perennial to 8′ with purple flowers much of the year. A house plant.

Divide plant in early spring. Pot in general-purpose potting soil.

Grow in a cool south window. Keep moist. Stand pot on wet pebbles. Fertilize every two months. Move outdoors in summer. Keep stem ends pinched.

MEXICAN ORANGE *Choisya ternata.* Evergreen shrub to 8′ with fan-like, yellowish-green foliage and fragrant, white flowers in late winter. Grows in warm climates.

Plant in spring or fall in average, well-drained soil. Grow in sun in cooler areas; in partial shade in hot areas. Water in dry weather. Fertilize in winter. After flowering, cut out branches that have grown too long. Spray with malathion to control mealybugs and scale.

MEXICAN TULIP POPPY *Hunnemannia fumariaefolia*, goldencup. Two-foot perennial commonly grown as an annual, with grey-green foliage and large yellow flowers. Grows almost everywhere.

Sow seeds outdoors where the plants are to grow a couple of weeks before last frost danger. Grow in full sun in average, well-drained, lightly limed soil. Fertilize when plants are thinned to 1′. Water only in dry weather and then sparingly.

MICHAELMAS DAISY *Aster,* hardy aster. One of the best perennial asters, to 6′, with fall flowers in many colors. Grows in temperate climates.

Grow like perennial asters (see aster). Divide clumps in spring or fall. Thin plants to about four strong shoots.

MICROMERIA One species called Yerba Buena. Creeping, spreading perennials to 6″ with mint-fragrant foliage and tiny, white or purple flowers. Used in rock gardens or as a groundcover. Grows in all but coldest climates.

Divide in spring or propagate by layering. Plant in shade in average, well-drained soil (about 2′ apart if used as a groundcover). Keep moist. Fertilize in spring.

MIGNONETTE *Reseda.* Annual to 15″ with sweet-scented flower spikes in red or yellow. Grows almost everywhere.

Sow seeds where plants are to grow when danger of frost is past. Make about two additional sowings at three-week intervals. Do not transplant. Grow in partial shade in average soil improved with a little extra humus. Water regularly. Fertilize when plants are thinned to 9″.

MINT *Mentha*, peppermint, spearmint. Perennials to 3′ with fragrant leaves. Grows almost everywhere.

Divide in spring and plant in average soil that contains humus. Grow in sun. Keep moist. A spring application of fertilizer will help a patch that is not growing well.

MIST FLOWER *Eupatorium coelestinum*, hardy ageratum, blue boneset. Two-foot perennial with dense clusters of blue or lavender flowers in late summer. Grows in temperate climates.

Divide in spring and plant in full sun in average, well-drained soil. Water moderately. Fertilize in spring.

MOCK ORANGE *Philadelphus*, sweet syringa. Deciduous shrubs to 12′ with white, often fragrant, spring flowers. Grows in all but extreme climates.

Propagate by seeds, layering or softwood cuttings in summer. Plant in spring or fall in sun in average, well-drained soil. Water in dry weather. Fertilize every 1–2 years in spring. Prune right after flowering.

MOMORDICA Balsam pear, balsam apple. A member of the gourd family grown as an annual vine to 20′. Produces pear-shaped or apple-shaped fruits. Grows best in warm climates.

Grow like gourds (which see).

MONKEY FLOWER *Mimulus.* Also identified as *diplacus.* Sticky-foliaged, evergreen shrubs to 5′ with a profusion of trumpet-shaped flowers in many colors. Grows mainly in California.

Plant the new hybrids in spring or fall in sun. Grow in average, well-drained soil. Water in dry weather. Fertilize every 1–2 years in winter. Prune after flowering.

MONKSHOOD *Aconitum*, aconite, wolfsbane. Perennials to 6′ with tall spikes of blue, purple, white or yellow flowers in late summer. Grows in temperate climates.

Divide plants in spring or fall. Grow in rich, humusy, well-drained soil in partial shade. Water regularly. Fertilize in spring. Stake tall plants. Burn diseased plants at once. Don't expect best bloom for 2–3 years.

MONSTERA Sometimes identified as *Philodendron pertusum.* Best known species (*M. deliciosa*) known as cerinam, Swiss cheese plant, Mexican breadfruit. Big evergreen climber to 20′ with huge leaves deeply lobed and perforated, calla-lily-like flowers, and edible fruits. A house plant; also grows outdoors in warmest climates.

Indoors. Grow like philodendron (which see). Propagate by cuttings of stem tips or by air-layering plants that have grown too large. Soil should contain double the usual amount of humus. Give a very sturdy support of bark-covered wood to climb up on.

Outdoors. Plant in spring or fall in filtered shade. The soil must be rich, humusy, well drained. Plant in the garden or a tub. Keep well watered. If planted in the garden, fertilize 2–3 times a year; if in a container, feed monthly with liquid fertilizer. Locate near the base of a tree it can climb on. Don't count on fruit. Prune after fruiting or flowering. Spray with phaltan to control leaf spot.

MONVILLEA Semi-erect, long-stemmed desert cacti to 10′ that flower at night. Grows in hot, dry climates.

For how to grow, see cactus, desert. Add a little extra humus to the standard cactus soil. Grow in a place that gets some shade.

MOONFLOWER *Calonyction aculeatum.* Sometimes called morning glory, to which it is related. Vine to 30′ with arrow-shaped leaves and fragrant white or purple flowers that open at night. Perennial in warm climates; treated as an annual elsewhere.

Grow like morning glory (which see).

MOONSEED *Menispermum canadense.* Perennial vine to 12′ with shiny leaves and bluish-black berries containing crescent-shaped seeds. Grows in all but very hot or dry climates.

This vine can easily become a pest so plant it in an out-of-the-way place. It makes a good groundcover. Propagate by spring-sown seeds, stem cuttings or suckers. Grow in average soil in sun or shade. Water in dry weather. Don't encourage it with fertilizer. Keep suckers cut out. Prune out dead wood in spring and cut back hard if necessary.

MORAEA Peacock iris. Plants growing from corms usually to 2′ or less, with iris-like flowers in many colors that last only a day. Grows in warm climates; also indoors.

Outdoors. Plant in the fall in light shade or sun. The soil should be of average quality, well drained. Set corms 4″ deep, 6″–9″ apart. Do not cut stems after flowering. Let plants die down naturally.

Indoors. Grow like freesia (which see).

MORNING GLORY *Ipomoea purpurea.* Annual vine to 15′ with blue, white or red flowers. Grows almost everywhere; also indoors.

Outdoors. Soak seeds in water for 24 hours or nick with a file before sowing. Start indoors 4–6 weeks before last frost, or sow outdoors where plants are to grow after frost. Morning glories like a light, sandy loam and full sun. Water regularly and fertilize once or twice after the little plants are established. Grow on wire mesh or a light wood trellis. Spray with malathion to control aphids.

Indoors. Sow seeds ½″ deep in pots of general-purpose potting soil in August. Thin to 2–3 plants to a pot. Grow outdoors in sun. Bring in before frost and place in a sunny, cool window. Water when soil feels dry. Fertilize every six weeks.

MOSS Spreading, low, flowerless, green plants of many kinds. Different mosses are found in different climates. Some also grow indoors.

Transplant in warm weather to the same general type of location and soil in which you found the plants. If bringing moss indoors, bring with it some of the soil in which it grew. In all cases, water the plants lightly when you move them. Thereafter, outdoor mosses should be able to take care of themselves (provided, of course, that you haven't planted them in alien situations). Indoor mosses should be given about the same amount of moisture they received outdoors.

MOUNTAIN ASH *Sorbus,* rowan tree, whitebeam, service tree. Deciduous trees to 60′ with flat-topped clusters of white spring flowers and red fall fruits. Grows best in cool, moist climates.

Plant in spring in the sun in average, well-drained soil. Water in dry weather. Fertilize in spring every 2–3 years. Prune in late winter. To control borers, spray trunk with malathion in May and twice more at two-week intervals.

MOUNTAIN HEATH *Phyllodoce.* Evergreen shrubs to 1′ with a profusion of purplish or yellow, bell-shaped flowers in spring or summer. Grows best in cool, moist climates.

Propagate by stem cuttings in August. Plant in spring or fall in partial shade. The soil must be well-drained and contain considerable humus. Water in dry weather. Fertilize every 2–3 years.

MOUNTAIN LAUREL *Kalmia latifolia.* Evergreen shrub to 10′ with large, pinkish-white flower clusters in spring. New varieties with deep pink to red flowers. Grows best in cool climates.

Plant in early spring or fall in sun or partial shade. Grow in light, well-drained, humusy soil that is strongly acid (pH of 4.5 to 5.5 is excellent). Mulch with oak leaves, pine needles or (less good) peatmoss. Water in very dry weather. Fertilize every 2–3 years. Cut back somewhat when transplanting, but leave it alone otherwise except to remove dead wood.

MULBERRY *Morus.* Deciduous trees to 60′ with sweetish, edible, red or white fruits. A popular variety is a small, weeping tree. The white mulberry grows from coldest climates southward.

Plant in spring or fall in average soil in a sunny spot. Water in very dry weather. Fertilize every 3–4 years. To be sure of fruiting, plant male and female specimens.

MULLEIN *Verbascum.* Biennials to 6′ with spikes of yellow, red, white or purple flowers. Grows in all but hot climates.

Divide in spring and plant in average soil that is well drained. Grow in the sun. Water in dry weather. Fertilize in spring.

MUSHROOM The type most commonly grown is *Agaricus campestris* (field mushroom), with a short, straight stem and large, smooth, white to ivory cap. Grows best under cover in high humidity at a temperature of 55°–65°.

Leave mushroom culture to the professionals unless you want to play with one of the small kits sold by mail and sometimes in garden supply stores.

MUSK MALLOW *Malva moschata,* musk rose. Perennial to 2′ with pink or white flowers. Grows in temperate climates.

Divide plants in spring or fall and plant in sun in average soil. Water in dry weather. Fertilize in spring.

MUSTARD GREENS *Brassica juncea.* Fast-growing, 10″ "greens" for boiling or salads. Grows almost everywhere, but most popular in the south.

Sow seeds in average soil in early spring and make succession sowings every week until a month before hot weather sets in. Then start making new sowings about August 15 and continue until a month before hard frost hits. (In other words, don't try to grow mustard greens in hot weather.) Thin plants to about 1′ apart in all directions. Water in dry weather. Keep cultivated. Cut the leaves when 4″ tall for salads and when 6″ for boiled greens. New leaves appear on cut plants.

MYRICA Different species called bayberry, wax myrtle, sweet gale, bog myrtle. Deciduous and evergreen shrubs or trees to 35′ with persistent but usually not evergreen foliage, sometimes aromatic, and in some species greyish, waxy, aromatic fruits used in making candles. Different species grow in different parts of the country.

Plant in spring or fall in a sunny

location. *M. pennsylvanica* and *M. californica* grow in poor, sandy soil (normally near the seacoast). *M. cerifera* requires a moist soil with large amounts of peatmoss. *M. gale* needs a very peaty, acid soil with a lot of moisture, since it normally grows in northern bogs. The last two, obviously, need to be well watered. Otherwise, the *myricas* are pretty easygoing.

MYRSINE AFRICANA African box. Evergreen shrub to 3′ with small, dark green leaves. Grows in warm parts of California.

Plant in spring or fall in partial shade or sun. Soil should be of average quality, well drained. Water in summer. Spray foliage with water occasionally. Fertilize every 2–3 years in spring.

MYRTLE *Myrtis communis.* The classic myrtle. Evergreen shrub to 10′ with glossy, aromatic leaves, white summer flowers and black berries. Grows in warmest climates; also indoors. See also periwinkle and Chilean quava.

Outdoors. Propagate by half-ripe stem cuttings. Plant in spring or fall in well-drained, slightly acid soil containing considerable peatmoss and sand. Mix in well-rotted or dried manure. Water abundantly during dry weather. Fertilize annually in spring. Prune in winter to shape.

Indoors. Plant in general-purpose potting soil containing extra peatmoss. Grow in a cool (60° or less) south window. Water when soil feels dry. Fertilize 3–4 times a year. Move outdoors in summer.

NANDINA DOMESTICA Sacred bamboo, heavenly bamboo. Evergreen shrub to 8′ with cane-like stems and large white flower clusters followed by red berries. Grows in warmest climates.

Plant two specimens if you want berries. Plant in spring or fall in sun. (*Nandina* also grows in shade but foliage doesn't turn such brilliant colors in fall.) Give a good, well-drained soil. Water regularly, especially if plant is in the sun. Fertilize annually in the spring. Cut some of the old canes to the ground in winter to promote most beautiful shape and keep plants fairly low.

NASTURTIUM *Tropaeolum.* Easy-to-grow annuals with bright orange, red, yellow and brown flowers. Some varieties are low and compact; others trailing or climbing to 12′. One climber is known as canary bird flower. Grows almost everywhere.

Sow seeds where plants are to grow when soil warms. Grow in full sun in average soil. Water in dry weather. Don't fertilize. Provide a light lattice for climbing types to grow on. Spray with malathion to control aphids, which can be very pesky.

NATAL PLUM *Carissa grandiflora.* Spiny, evergreen shrub to 15′ with glossy foliage, fragrant white flowers and edible red fruits. Grows in subtropics; also indoors.

Propagate by layering of notched branches. Plant in spring or fall in sun in average, well-drained soil or general-purpose potting soil. Water deeply every few weeks during dry seasons. Fertilize every spring. Move house plants outdoors in summer. Prune old growth in winter to promote fruiting, which takes place on new growth.

NAUTILOCALYX Plants to 2′ with fine leaves and small yellow flowers. A house plant.

Grow like *episcia* (which see). Propagate by stem cuttings. Give plenty of warmth. Pinch center shoots to promote bushiness, and cut out old stems occasionally.

NECKLACE VINE *Muehlenbeckia complexa*, mattress vine, maidenhair vine, wire vine. Twining vine to 25′ with small, round leaves and white flowers. Mainly a house plant, but sometimes grown outdoors in warmest climates.

Propagate by stem cuttings. Pot in general-purpose potting soil. Grow in a south window. Water when soil feels dry. Fertilize every 2–3 months. Move outdoors in summer. Prune rather severely when bringing plant back indoors in the fall. Grow as a trailer or up strings or a light trellis.

NECTARINE *Amygdalus persica*. Deciduous tree to 25′ bearing small, smooth-skinned fruits very much like peaches. Grows best in California.

Grow like peach (which see).

NEILLIA Deciduous, spirea-like shrubs to 6′ with clusters of pink, tube-shaped flowers in spring. Grows in all but coldest climates.

Grow like spirea (which see). Prune after flowering.

NEMESIA Compact annual to 2′ with showy flowers in many colors. Grows best in cool climates.

Sow seeds indoors 8–10 weeks before last frost. Keep in a cool, sunny window. Move outdoors when weather is reliable and plant in sun in average soil improved with humus. Keep soil moist. Fertilize when plants are established and again about a month later. Pinch stem ends to promote bushy growth.

NEMOPHILA MENZIESI Baby blue eyes. Six-inch annual with bell-shaped, blue flowers. Grows in cool climates.

Sow seeds in early spring in average garden soil. Grow in sun or partial shade. Fertilize when plants are thinned to 6″. Water in dry weather.

NEOREGELIA Popular species are painted fingernail, marble plant, crimson cup. Bromeliads to 18″ with large, often beautifully colored leaves and some with showy bracts and flowers. Grows indoors; also in subtropics.

Grow like *aechmea* (which see). *Neoregelias* need a couple of hours of sun every day to look their best.

NEPETA Perennials of various types and sizes, mostly with blue flowers. One species called catnip, catmint. Another called ground ivy, gill-over-the-ground, field balm. Grows in temperate climates. Ground ivy also grows indoors.

Outdoors. Propagate from spring-sown seeds or summer cuttings. Grow in light shade or sun in average soil. Water in dry weather. Fertilize larger types in spring.

Indoors. Grow ground ivy in general-purpose potting soil in an east or west window. Water when the soil feels dry.

NEPHTHYTIS AFZELI Vine growing from a rhizome to 10′, with arrow-shaped leaves. Mainly a house plant.

Propagate by stem cuttings or division of the rhizomes. Pot in general-purpose potting soil. Keep moist and warm. Grow in an east or north window. Give a little plant food every 2–3 months. Provide a bark-covered stake for plant to grow on. Nephthytis also does well in water alone.

NERINE One species called Guernsey lily. Bulbous plant to 3′ with spidery fall flowers in reds or orange and leaves coming later. Grows in warm climates.

Plant bulbs 3″ deep and 6″ apart in early fall in good, sandy soil in the sun. Water well until flowers die; then reduce supply somewhat until May, when leaves turn yellow; then withhold it almost entirely. Fertilize in early fall.

NEW ZEALAND FLAX *Phormium tenax*. Evergreen perennial to 15′ with long, sword-shaped, variously-colored leaves and red or yellow spring flowers. Grows in warmest climates.

Divide in spring and plant in sun in average soil. Water until established. Then you can pretty much forget the plant except to keep it from spreading too far.

NICOTIANA Flowering tobacco. Two-foot annual with fragrant flowers mainly in white, pinks and red. Grows almost everywhere.

Sow seeds indoors eight weeks before last frost. Plant in full sun in average soil when weather is warm. Water regularly. Fertilize when little plants are established.

NIDULARIUM Tree-perching bromeliads to about 1′ with leaves in a rosette more or less surrounding red bracts. Grows outdoors in subtropics; also indoors.

Grow like *aechmea* (which see). *Nidularium* requires plenty of heat in summer—75° at night and more in the day. In winter, nighttime temperature can drop to about 60°.

NIEREMBERGIA Cup flower. Sprawling perennials treated as annuals; usually only 8″ tall (but one species much taller) and covered with cup-shaped, white or lilac summer flowers. Grows almost everywhere.

Sow seeds indoors 8–10 weeks before last frost; shift outdoors into humusy soil. Grow in partial shade. Water regularly. Fertilize when young plants are established and again in late June. In warm climates, plants will survive the winter.

NIGELLA Love-in-a-mist, devil-in-the-bush. Annual to 18″ with lacy leaves and blue or white flowers. Grows almost everywhere.

Sow seeds outdoors as soon as soil can be worked in the spring. Grow in average soil in a sunny spot. Water in dry weather. Fertilize when plants are thinned to 8″.

NIGHTBLOOMING CEREUS *Hylocereus undatus*. A climbing desert cactus to 40′ with snake-like stems and huge, white, funnel-shaped flowers that open at night in summer and fall. Grows in subtropics; also indoors.

Outdoors. Propagate by stem cuttings which are allowed to dry in the air and form a callus before they are planted shallowly in sand. Don't expect bloom for several years. Plant in the sun in two parts loam, two parts sand and one part humus. The soil must be well drained. Mix in a little bone meal. In spring and summer, water thoroughly whenever soil feels dry; but little water is required at other times. Fertilize lightly in early spring and twice after that at monthly intervals. Grow on a sturdy trellis to which stems are tied in order to prevent them from tangling. Cut out damaged parts when you find them and dust wounds with sulfur.

Indoors. Pot in same soil mixture and grow as above. During winter, water just enough to keep plant from becoming desiccated. Temperature in the low 60's is best at this time of year. Keep in a south window at all times.

NORFOLK ISLAND PINE *Araucaria excelsa.* Very symmetrical evergreen tree to 10′ (and much more outdoors) with neatly tiered branches. Mainly a house plant, but sometimes grown outdoors in subtropics.

Pot in general-purpose potting soil. Water when soil feels dry. Fertilize about three times a year. Grow in a north window at no more than 60°. Move outdoors into light shade in summer.

OAK *Quercus.* Favorite deciduous and evergreen trees to 100′. There are few parts of this country where one or more species does not grow.

Plant in spring or fall in sun in average soil or better. Water in dry weather. Most oaks need considerable moisture. Fertilize every 3–5 years. Keep dead wood cut out. Spray with one part liquid lime-sulfur and nine parts water to prevent leaf blister, which causes yellowish raised spots on upper surfaces of leaves. Make application in early spring before buds open. Spray 2–3 times at two-week intervals with Bordeaux mixture to control anthracnose, which causes expanding brown spots on leaves in wet weather. Make first application as buds open.

Note. In California and some other areas where live oaks thrive, paved terraces are often established under the trees. This apparently endangers the life of the trees and is therefore not a recommended practice.

OCHNA MULTIFLORA Broadleaved evergreen shrub to 6′ with yellow flowers that develop red centers and then five small green fruits which turn black. Grows in southern California.

Propagate by half-ripe stem cuttings in summer. Plant in spring or fall in partial shade in acid, humusy, well-drained soil. Water in summer. Fertilize in spring. Prune in winter.

OCONEE BELLS *Shortia galacifolia.* Evergreen, creeping perennial with round leaves just above the soil and white flowers 8″ tall. Grows in the middle south.

Propagate by stem cuttings in early summer. Plant in fall in partial shade in slightly acid, humusy, well-drained soil. Keep moist. Fertilize occasionally.

OENOTHERA Evening primrose, Missouri primrose, sundrops. Biennials and perennials to 5′ with poppy-like, yellow, white or rose flowers that open during the night or day, depending on the species. Grows almost everywhere.

Divide plants in spring. Oenotheras need sun and an average, well-drained soil containing a little extra humus. Keep watered and cultivated. From early spring into midsummer fertilize at six-week intervals. Mulch in winter.

OKRA *Hibiscus esculentus,* gumbo. Vegetable to 5′ with small green pods full of seeds and a mucilage-like liquid. Grows best in the south.

Okra is not particular about soil but likes it sweet. Sow seeds 1″ deep in rows 3′ apart when danger of frost is past. Thin plants to 18″–24″, depending on whether variety is small or large. Give full sun. Water in very dry weather. Fertilize once or twice with general-purpose plant food with a low nitrogen content. Pick pods when small.

OLD MAN CACTUS *Cephalocereus Senilis,* Mexican old man, golden old man, Peruvian old man, old man of the Andes. Large (to 50′), columnar desert cacti of several genera. They are distinguished by the profuse hair on the stems. Grows in hot, dry climates.

For how to grow, see cactus, desert.

OLEANDER *Nerium oleander,* rose bay. Poisonous evergreen shrub to 25′ with showy white, red, pink or purple flowers in spring and summer. Grows in warmest climates; also indoors. See also yellow oleander.

Outdoors. Propagate by layering or by semi-hardwood cuttings taken in winter or early spring. Plant in spring or fall in full sun. Give a well-drained, average soil and fertilize in late winter or early spring. Keep cultivated. Water regularly until flowers fade. Prune after flowering. Old stems can be cut back hard (flowers come on new wood). Spray with dormant oil in spring to control scale insects.

Indoors. Grow in general-purpose potting soil. Repot in March every year and water whenever soil feels dry. Fertilize lightly every two months. Keep in a south window in a cool room. Move outdoors when weather is reliably warm and keep in a sunny spot. In the fall, cut back branches about 25% and bring indoors. Water sparingly and do not feed until growth starts again.

OLIVE *Olea europaea.* Handsome evergreen tree to 25′ bearing blackish fruits which are used for pickles and oil. Grows best in hottest parts of California. Two varieties are needed to insure fruiting.

Plant in spring in a sunny location in well-drained, average soil. Space trees 30′ apart. Water deeply in winter and until fruit reaches full size. Fertilize in spring. In winter prune out all branches below 3′ and undesirable branches above that until trees start bearing at the age of about five years. Thereafter, thin out tops, removing the small branches which bore fruit the previous summer. Thinning the fruits when they are small helps to produce larger fruit. To control scale, spray in winter with oil emulsion.

ONION *Allium cepa.* Onions are grown either to green scallion size or to produce mature white, yellow or red bulbs. They are grown from sets (small dried onions), seedlings or seeds almost everywhere except in desert regions.

Onions require full sun; more or less constant moisture; regular light cultivation to keep down weeds. The soil should be rich, well-drained, deeply dug and reasonably free of stones. To assure necessary fertility, mix into the soil an ample amount of humus and general-purpose fertilizer and lime. Sidedress plants that are to be grown to maturity with general-purpose fertilizer once or twice during the growing season.

Scallions are harvested while they are young and bright green. Mature onions are harvested in the fall when the tops turn yellow (it takes four months or more to reach maturity). Cut off tops 1″ above the bulbs and store in a cool, dry, airy place.

Do not try to raise scallions and mature onions in the same row, because when you pull the scallions you may disturb the roots of those you want to grow on. Pulling weeds near onions also endangers plants.

Onions from sets. This is the easiest way to grow onions. Plant the sets ½″ deep as soon as the soil can be worked in the spring. Space sets 3″–4″ apart if they are to be grown to mature bulbs; 1″ apart if they are to be eaten as scallions. Allow 15″–18″ between rows. If sets are to be grown to maturity, plant only those that are less than ⅝″ diameter; larger ones will go to seed long before reaching full size.

Onions from plants. Seedling plants may be purchased from growers in your area. They are used to produce mature onions. Plant like sets 10–14 days later in the spring.

Onions from seeds. Sow seeds ¼″ deep in rows 15″–18″ apart as soon as the soil can be worked. Thin to 3″.

OPUNTIA Different species called prickly pear, cholla, Indian fig, bunny ears, beavertail, dominoes, Joseph's coat, grizzly bear cactus, boxing gloves, Mexican dwarf tree. Desert cacti to 18′, variously shaped, with spines, tufts of barbed hairs and stalkless yellow, red or white flowers. Grows mainly in hot, dry climates but some species grow quite far north. Also a house plant.

For how to grow, see cactus, desert. The soil for opuntias should contain somewhat more sand than most cacti like. Water moderately and fertilize sparingly even while plants are growing their best. If kept dry in fall and winter, plants need little protection against even very cold weather.

ORANGE *Citrus Sinensis.* Evergreen trees to 25′ with fragrant, white blossoms and sweet orange fruits. Grows in warmest climates. The otaheite and calamondin orange—small trees with inedible fruit—grow indoors.

Outdoors. Plant in February in a sunny location. The soil should be a deep, well-drained, sandy loam. Space trees 30′ apart. Water deeply but only when soil becomes dry far down and when leaves show signs of wilting. Fertilize three times a year with special citrus food (such as 5-7-5, 6-4-6 or 6-6-6)—in early February, early June, and early November. Apply at the rate of 1 lb. per year of age up to a maximum of 10 lb. It is advisable to grow a cover crop under the trees in summer and plow it under. Trees in limestone soil should also be given a nutritional spray of copper, zinc and manganese in January.

Prune trees in mid-spring or summer. Cut out dead wood, low-hanging branches, suckers and watersprouts. The main aim of pruning is to keep the tree shaped well, and somewhat open in the middle.

To protect newly planted trees, mound soil up around the trunks to a height of 1′ or more in early

November. Remove this as soon as weather is reliably warm after the first of the year. Heat and air movement are necessary to protect large trees against killing frosts.

To control scale, spray in June with an oil emulsion spray. Dusting with sulfur after blooming helps to control rust mites, though these do not actually harm the fruit inside.

Indoors. Orange trees are easily grown from seeds taken from store fruits and planted in general-purpose potting soil; but they will not flower or fruit. Better buy plants. Grow in general-purpose potting soil in a south window at a daytime temperature no higher than 70° and a nighttime temperature of 45°–55°. Keep soil moist except in fall and early winter, when it should be on the dry side. Spray foliage with water frequently. Fertilize every two months when plant is making growth. Move outdoors in summer. If necessary to encourage fruiting, the powdery pollen can be transferred with a cotton swab from one flower to the moist, sticky stigma of another.

ORANGE BROWALLIA *Streptosolen jamesoni.* Evergreen shrub to 6′ with clusters of orange flowers off and on throughout the year. Can be trained as a vine. Mainly a house plant.

Pot in general-purpose potting soil. Keep moist. Fertilize 3–4 times a year. Keep in a cool south window. Pinch stem ends to promote branching, and prune rather severely after flowering. Train to a light lattice.

ORANGE JASMINE *Muraya* (or *murraea*) *exotica.* Subtropical shrub to 10′ with glossy, evergreen leaves and fragrant white flowers and red berries.

Propagate by stem cuttings. Plant in spring or fall in partial shade in average, well-drained soil. Water in dry spells. Fertilize about February and again in May. Prune when flowering stops in late summer or fall. Apply chelated iron to soil if leaves turn yellow between the veins.

ORCHID Huge tribe of showy flowers. They are either epiphytic (tree-perching) or terrestrial (growing in soil). The epiphytes are tropical and mainly grown indoors. Some of the terrestrials are also tropical and grown indoors, but others are hardy and grown outdoors.

Note. The culture of orchids varies. Instructions given here are generally applicable, but there are many exceptions to the rules. See entries for specific genera listed in italics.

Epiphytic orchids indoors (*cattleya, coelogyne cristata, dendrobium, epidendrum, lycaste skinneri, vanda, zygopetalum*). Pot or repot plants after flowering in osmunda fiber or chopped bark. Use special clay orchid pots and fill one-third with shards. The base of the pseudobulb or rhizome should be about level with the top of the pot so water drains away from it. Spread out roots and pack the fiber or bark very firmly around them. Water, but not too much, until growth starts.

The general rule for watering orchids is to use non-alkaline water (rainwater is excellent) and to apply it only when the osmunda or bark dries out. Then pour several quarts of water through the "soil."

Fertilize plants no more than

every four weeks while they are in active growth, but not while they are resting after flowering. Use liquid orchid food according to the manufacturer's directions. If potted in chopped bark, make occasional additional doses of straight nitrogen. (Yellowing of foliage indicates that you may be feeding plants too much.) Orchids should be grown in well-ventilated but not drafty rooms. To provide the humidity that is essential, keep pots standing on wet pebbles at all times. In addition, while plants are making active growth, spray them daily with water (provided the foliage dries rather rapidly). While plants are resting, spraying should be reduced.

Orchids need a normal daytime house temperature, but a nighttime temperature of about 60°. In the north, grow plants in a south window which should be screened with cheesecloth at midday during spring and summer. In the south, grow in a south window that is screened at midday all year. If you wish, move plants outdoors in summer provided the temperature does not fall below 60°. Start them in the shade; then gradually move them into a place where they get morning sun but no more than filtered sun later in the day.

If large, hard, brown-black spots form on orchid leaves, reduce the amount of sun the plants are getting. If the brown-black spots are soft, however, cut them out and treat the wounds with Wilson's Anti-Damp.

To control scale, scrub plants with soap and water or spray several times with malathion.

In cutting orchid flowers, take care to remove as much of the flower sheath as possible; otherwise, water may be trapped by the sheath and the immediate area will be attacked by fungus.

Terrestrial orchids indoors (*calanthe, cymbidium, cypripedium*). These are handled as above but potted in different soil mediums. See entries for specific genera.

Hardy terrestrial orchids outdoors (*bletilla, habernaria, ladyslipper, orchis*). As a general rule, plant in spring or fall in well-drained average soil containing considerable peatmoss. Grow in partial shade. Keep moist. Mulch in fall after ground freezes.

ORCHID CACTUS *Epiphyllum*. Spineless jungle cacti with flat, jointed stems up to 20′ long and with spectacular flowers in red, yellow or white. Grows indoors; also outdoors in subtropics.

Plant in equal parts of loam, humus and sand. Be sure drainage is good. Grow in an east or west window or partial shade outdoors. Protect from frost in winter and water very sparingly. In spring and summer, water when the soil feels dry and feed every three weeks. After buds develop, spray foliage with water in early morning; this helps to stop bud-drop. House plants should be moved outdoors in summer.

ORCHIS Perennial wildflower to 1′ with basal leaves and terminal clusters of white or purple flowers. Grows mainly in the east.

Grow like *habernaria* (which see).

OREGANO *Origanum vulgare*, pot marjoram, wild marjoram, wintersweet. Perennial to 2′ with pinkish flower spikes, grown for the leaves which are used for flavoring. Grows in temperate climates.

Divide in spring or fall and plant in sun in average soil. Water in dry weather.. Fertilize in spring.

OREGON GRAPE HOLLY *Mahonia aquifolium*. Evergreen shrub to 10′ with holly-like leaves that turn bronze-red in fall, yellow flower clusters in spring and blue fruits in summer. Grows in mild climates.

Propagate by layering. Plant in spring or fall in a lightly shaded location not exposed to cold winds or too much winter sun. A north or north-east exposure is usually good. Grow in a well-drained but moist, average soil containing some extra humus. Water in dry spells. Fertilize every 1–2 years in spring. Prune after fruiting to control growth and shape. Cut out woody stems that stick out above the crown of foliage.

OSAGE ORANGE *Maclura pomifera*, bow-wood. Spiny, deciduous tree to 50′ with greenish, orange-like, inedible fruits. Grows in mild and warm climates.

Plant in early spring or fall in sun. Grow in average soil. Water in very dry weather. Fertilize every 2–4 years.

If used in a hedge, young plants should be spaced 6″ apart. Shear the tops in summer, when the plants are making good growth, to force side branches. Trim to shape later in the year. This procedure should be followed every year. Trim plants so they are wider at the bottom than at the top.

OSMUNDA Different species called cinnamon fern, royal fern, interrupted fern. Large ferns with erect, coarse fronds to 6′ arranged in clusters. Grows mainly in temperate climates although royal fern is also found in subtropics.

For how to grow, see fern. Grows best in moist, humusy soil in light shade. Also grows in sun.

OSTRICH FERN *Pteretis nodulosa*. Coarse fern with fronds to 8′ and arranged in the shape of a vase. Grows mainly in the east.

For how to grow, seen fern.

OXALIS Wood sorrel. Bulbous plants to 6″ with clover-like leaves and variously colored flowers. One or more varieties grows in almost every region. Some are good house plants.

Outdoors. Plant bulbs 2″ deep and 4″ apart in average soil in sun or shade in the spring. Water in very dry weather. Fertilize every 2–3 years. In cold climates, treat like *gladiolus* (which see).

Indoors. Pot several bulbs in a large pot filled with general-purpose soil in the fall. Set the bulbs 1″ deep and 1″ apart. Place in a 60° south window and water very little until growth appears; then increase water. After flowers die, reduce water gradually. Then store bulbs in dry soil in their pot in a cool place until next fall.

OXERA PULCHELLA Royal climber. Evergreen vine to 10′ with large clusters of creamy-white, trumpet-shaped flowers in summer. Grows in southern California.

Propagate by seeds or stem cuttings. Plant in spring or fall in light, well-drained, humusy soil. Grow in a spot protected from hot midday sun. Keep soil evenly moist. Fertilize in spring and fall. Tie stems to a support. Prune lightly right after flowering.

OYSTER PLANT *Tragopogon porrifolius,* salsify. Plant to 3′ with several fleshy, white roots that taste something like oysters. Grows in cool climates.

Grow like parsnip (which see).

PACHISTIMA Different species called rat-stripper, myrtle boxleaf, Oregon boxwood. Evergreen shrubs to 4′ with small glossy leaves and inconspicuous flowers. Grows in mild climates.

Propagate by layering. Plant in spring or fall in sun or partial shade. Grow in average, well-drained soil. Water in dry weather. Fertilize every 2–3 years. Prune after flowering; plants can be sheared.

PACHYCEREUS Different species called Mexican giant cactus, organ pipe cactus, Indian comb cactus, hair-brush cactus. Large, tree-like, desert cacti to 50′, shaped like candelabra. Grows in hot dry climates.

For how to grow, see cactus, desert. Plant in soil with a little extra humus. Water well when making growth in spring and summer.

PACHYSANDRA Evergreen plants to 8″ grown as a groundcover in all but most extreme climates.

Propagate by cuttings or divisions in spring. Plant in average soil in light shade. Water in very dry weather. Fertilize in spring if you want plants to grow faster.

Extremely thick plantings of pachysandra are sometimes killed by fungus; therefore it pays to thin them out occasionally. If plantings grow far beyond their borders, a good way to move the plants is to handle like long rolls of sod.

PALM Subtropical trees, sometimes very tall, sometimes shrubby, with open fronds. Some genera grow indoors.

Outdoors. Plant in spring in a sunny location (however, very small specimens in pots should be started in partial shade and gradually accustomed to full sun). The soil should be very well drained and neutral or slightly acid. A mixture of two parts loam, two parts humus and one part sand is about right. Do not let soil dry out completely at any time. Fertilize in late winter or early spring, and in midsummer. In warmest climates, a third feeding in the fall is recommended. Large trees can use about 10 lb. at a feeding. Remove fronds when they die; also cut away old stalks after production of flowers and fruits. If cluster palms get out of hand, some of the stems may be cut off at the base. But never damage or remove the terminal growth.

Spray with malathion to control scale and other insects.

Indoors. Pot in soil mixture above. Make sure pot drainage is good. When repotting every other year, roughen up the roots slightly so they will spread into the new soil. Keep soil moist at all times. Fertilize with liquid plant food at three-week intervals while tree is making growth.

As a rule, grow most palms in an east window at a temperature of 65°. Spray with malathion to control red spiders, which cause leaves to turn brownish. To control aphids and mealybugs, spray foliage with water once a week.

PALMETTO *Sabal.* Sizable genus of palm trees, one like a small shrub with no visible trunk, others to 90′. All have fan-shaped palms. Grows widely in warm climates.

For how to grow, see palm.

PAMPAS GRASS *Cortaderia selloana.* Giant grass to 20′ with showy plumes in late summer. Grows in warmest climates.

Divide in spring and plant in sun in average soil. Water to get established. The grass needs little attention thereafter. Watch out, however, that it doesn't spread too far.

PANDANUS Screw pine. Shrub to 3′ (taller outdoors) with long, spiny-edged, white-banded leaves and roots that push out of the ground. Mainly a house plant.

Propagate by suckers in February. Pot in general-purpose potting soil. Grow in an east or west window. Water regularly and spray leaves with water from January through August. Fertilize 2–3 times during this period. Keep warm but ventilate the room in which plant is growing.

PANDOREA Different species called wonga-wonga vine, queen of Sheba, bower plant. Evergreen vines to 30′ with pink or white flowers in summer. Grows in warmest climates.

Propagate by cuttings of green stems. Plant in rich soil with plenty of humus. Grow in partial shade out of the wind. Keep soil moist and fertilize in spring and fall. Train on a lattice. Cut out dead and damaged wood in early spring. Spray with malathion to control aphids.

PANSY *Viola tricolor hortensis,* heart's ease, Johnny jump-up. A biennial to 8″ producing beautiful spring flowers in many colors. Grows almost everywhere.

Buy freshly harvested seed and sow outdoors in August. The seedbed soil should be a well-pulverized mixture of one part loam, one part humus and one part sand. Mix in general-purpose fertilizer before seeds are sown. Keep soil moist. Shade the seedbed until seeds germinate, then shade only at midday. Transplant seedlings, when large enough to handle, to 6″ apart. They may be in a coldframe, a nursery bed or the garden where they are to bloom. The plants should now have full sun. Keep moist.

When soil freezes, mulch plants by placing salt hay around them and covering lightly with evergreen boughs or a little salt hay. In early spring, move plants to their final location (if they are not already there). Grow in rich, humusy soil which is kept moist. Keep blossoms picked off.

PAPAYA *Carica papaya,* pawpaw, melon tree. Tree to 25′ producing edible, yellow, melon-like fruits. Grows only in subtropics.

Plant a tree that is guaranteed to be bisexual; otherwise, plant both male and female specimens. Plant 10′ apart in spring in full sun. The soil must be rich, humusy, moisture-retentive, but well drained. Keep cultivated. Water often in dry weather. Fertilize every two weeks with a cupful of general-purpose plant food. Spray with rotenone to control spider mites; with zineb to prevent fungous diseases. When fruits form, enclose them in paper or plastic bags to keep off fruit flies. Trees need protection against freezing temperatures.

PAPER MULBERRY *Broussonetia papyrifera.* Deciduous trees to 50′ with handsome foliage. Grows in mild climates.

Plant in early spring or fall in a sunny location in average soil. Water in very long dry spells. Fertilize every 3–5 years, Actually, the tree needs very little attention. It even grows well in cities.

PARKINSONIA One species known as Jerusalem thorn, horse bean, ratama. Evergreen trees to 25′ with long spines and handsome yellow flowers. Grows in warmest climates.

Plant in spring or fall in sun in average soil. The trees almost never need watering. Fertilize every 3–4 years in early spring.

PARROT BEAK *Clianthus puniceus.* Evergreen vine to 12′ with open foliage and red, sweet-pea-like flowers with parrot beaks. Grows in warmest climates.

Propagate by stem cuttings in spring. Plant in spring or fall in a sunny location protected from wind. The soil should be well drained and contain a fair amount of humus and sand. Water regularly when plant is blooming in spring; also in dry weather. Fertilize in early spring. Tie stems to a light trellis. Prune out dead and weak wood in early spring.

PARSLEY *Petroselinum.* Biennial herb to 15″, often grown as an annual. Grows almost everywhere.

In early spring sow seeds in average soil in rows 15″ apart. Give full sun. Water in dry weather. Fertilize lightly when plants are thinned to 6″. In the fall, after soaking the soil, dig up some of the plants with balls of earth and set in pots. Cut back to the crowns. Water sparingly until growth starts. Grow in a cool, south window. Keep older leaves picked off.

PARSNIP *Pastinaca sativa.* Vegetable to 3′ with long, white roots eaten in fall and winter. Grows in cool and cold climates.

The soil for parsnips should be deeply dug and well pulverized, cleared of stones and enriched with humus and fertilizer. Plant seeds in early spring in rows 18″–24″ apart. Thin plants to about 4″. Keep weeded and water in dry weather. Fertilize when plants are about 6–8 weeks old. Roots are ready to be dug in the fall but can be left in the ground all winter and dug at any time.

PARTRIDGEBERRY *Mitchella repens,* squawberry, teaberry, twinberry. Creeping evergreen only a few inches high with red berries. Grows in the east.

Dig up rooted sections with as much soil as possible or propagate by layering. Plant in early spring or late fall in shade in well-drained, slightly acid, humusy soil. Water in dry weather.

PASSION FLOWER *Passiflora,* passion fruit. Genus of evergreen vines to 35′ with intricate, multi-colored summer flowers. One species called granadilla has large, brownish-purple, edible fruits. Another called Maypop produces small, yellow, edible fruits. All species grow in warmest climates. Some also grow indoors.

Outdoors. Propagate by layering. Plant in spring or fall in average, well-drained soil which should be kept moist. Fertilize monthly in spring and summer. The vine needs sun but should be protected at midday in summer. Spray foliage with water often. Grow on a sturdy lattice. Keep thinned out. Cut back severely in late fall after second year. Spray with rotenone to control insect pests.

Indoors. Pot in general-purpose potting soil and keep in a south window during winter but move outdoors in summer. Don't water too much until growth starts in the spring, then water and fertilize regularly. Spray foliage with water. Cut back hard in the fall when you bring vine indoors.

PATIENCE *Impatiens.* Annual to 3′ with pink, red, white or purple flowers. Grows almost everywhere; also indoors.

Outdoors. Sow seeds indoors 8–10 weeks before last frost. Shift into partial shade in well-drained, average soil. Keep watered. Fertilize when young plants are established and once again in the summer. Cuttings taken from these can be grown indoors the following winter.

Indoors. Propagate from seeds sown outdoors in May or June or by stem cuttings taken in June or July. (Note that plants from seeds do not always come true; so if you have a favorite variety, you'd better propagate by stem cuttings.) Plant in general-purpose potting soil and grow in the shade until fall. Then move indoors and grow in a south window. Water when soil feels dry. Fertilize 2–3 times during the fall and winter. Don't carry over plants.

PAULLINIA THALICTRIFOLIA Evergreen vine to 10′ with lacy, fern-like leaves. Subtropical.

Propagate by stem cuttings in early spring. Plant in spring or fall in a tub filled with general-purpose potting soil with extra humus. Keep in a lightly shaded location away from wind. Water regularly. Fertilize in early spring with general-purpose plant food rich in nitrogen. Grow on a light lattice. Pinch stem ends often to promote bushy growth. Spray with malathion to control aphids.

PAULOWNIA TOMENTOSA Empress tree. Deciduous tree to 50′ with large leaves and violet flowers appearing in spring before the leaves. Grows in warm and mild climates.

Plant in spring or fall in a sunny, sheltered situation. The soil should be light, deep and well drained. Water in dry weather. Fertilize every 2–3 years in early spring. If tree is killed to the ground by cold, it will probably send up new shoots, of which all but one should be cut out.

PEA *Pisum sativum.* Spring vegetable which grows almost everywhere in cool weather. Dwarf (down to 14″) and tall (up to 60″) varieties are available. Wrinkle-seeded varieties are of better quality than the smooth-seeded but cannot be planted quite so early in the year.

Peas grow in any average, well-drained soil that is sweet and enriched with fertilizer. Full sun is required. As soon as the frost is out of the ground and the soil can be worked, sow the seeds 2″ deep in rows 18″–24″ apart. Thin plants to 2″–3″. Make several succession sowings of dwarf varieties, provided they will mature before hot weather sets in. Water if the spring is unusually dry. Support plants on twiggy branches, chicken wire or strings strung between posts. Pick peas just before they are to be eaten. Do not plant them in the same spot two years in a row.

PEACH *Amygdalus persica.* Deciduous tree to 25′ with pink spring flowers and familiar fruits. Grows almost everywhere but does best where temperatures do not dip too far below zero.

Plant one-year-old trees in early spring (or in late fall in warm climates) in a high, sunny location. The soil should be a deep, well-drained, sandy loam containing some humus which holds moisture, and lime. Space trees 20′–25′ apart. Scatter general-purpose fertilizer on the ground around each tree in early spring at the rate of 1 lb. per year of age up to a maximum of 10 lb. Water deeply in dry spells. For best fruit, keep soil cultivated.

Peach trees should be pruned in late winter so that they are rather open in the center and shaped more or less like a bowl. At planting time, cut the branches back to 1–2 buds. The next year, remove all but 3–4 strong, out-reaching branches which are evenly spaced around the trunk and up and down it. Cut off the leader just above the top branch. Cut back the branches you have saved to the first bud or side branch. Keep all branches equal length. Thereafter, prune trees lightly every year to remove dead wood, suckers, low-hanging limbs. Don't allow new main branches to develop, and don't remove too many of the small growths. After the tree is bearing well, however, pruning should be more severe.

To control insects and disease, spray as follows: In late winter before buds swell, apply dormant lime-sulfur spray. Then apply all-purpose fruit tree spray immediately before flowers open, at time petals drop, ten days later, ten days after that, ten days after that, three weeks before fruit is ready for picking. In addition, if borers become a problem, spray the trunk 2–3 times in late spring with malathion. Burn fallen leaves and do not leave old fruits hanging on trees.

PEANUT *Arachis hypogaea,* goober. Sprawling annual to 18″ grown in the south. The nuts are suspended from the branches but are produced underground.

Peanuts grow in relatively poor soil—a sandy loam containing some humus. The soil should be limed and enriched with superphosphate and muriate of potash. After last frost, remove seeds from pods and plant 1″ deep and 1′ apart in rows 3′ apart. Water in dry weather. Cultivate very carefully as the plants grow, being careful not to harm the roots; then stop cultivating after the flowers have buried themselves in the soil. Mulching helps to keep down weeds and makes harvesting of crop easier. Four months are required for the peanuts to mature. Then pull up the plants or, if the ground is hard, dig up the nuts with a fork. Store in a dry place.

PEANUT CACTUS *Chamaecereus silvestris.* Three-inch desert cactus with a cluster of peanut-shaped, spiny stems and red spring flowers. Grows in hot, dry climates; also indoors.

For how to grow, see cactus, desert.

PEAR *Pyrus.* Deciduous tree to 40′ with white spring flowers and fruits of various shapes and colors in summer and fall. Grows best in cool, moist climates.

Plant two varieties of one-year-old trees in spring in a high, sunny location in deep, well-drained, loamy soil. Space trees 20′–25′ apart. Fertilize in spring with only 2 or 3 cups of ammonium nitrate. Water well in dry weather. To control insects and disease, follow schedule for apples (which see).

Cut pear trees back one-half at planting time. Thereafter prune in late winter. In the second year, remove all but 3–4 branches which are evenly distributed around the trunk and spaced 8″ apart up and down. From then on until tree is bearing, thin out excess branches and head back others lightly to form a compact tree. After tree is bearing, prune lightly every year to eliminate excess food and maintain a good supply of the spurs on which the fruit is borne. In spring after fruit is set, if the amount of fruit seems excessive, reduce the clusters to one or two fruits each.

PEARL BUSH *Exchorda racemosa.* Deciduous shrub to 10′ with pearl-like buds followed by white spring flowers. Grows in mild and warm climates.

Propagate by layering. Plant in spring or fall in a sunny spot in average, well-drained soil. Water in dry spells. Fertilize every 2–3 years. Prune after flowering.

PEA TREE *Caragana arborescens.* Deciduous shrub or tree to 20′ with pea-like, yellow flowers in spring. Grows in cold and cool climates.

Plant in early spring in full sun in sandy soil. Water in dry weather. Fertilize every 2–3 years. Prune after flowering.

PECAN *Carya pecan.* Deciduous tree to 120′ with valuable nuts. Fruits reliably only in the south and does best if two different varieties are planted.

Plant only those varieties recommended for your particular area. Planting is done in early spring. A sunny location is required. Allow plenty of space around each tree because in time its roots will spread 30′–40′ in all directions. Plant in average, well-drained soil. Apply 2–3 lb. of fertilizer per inch of trunk diameter at planting time and each year thereafter. For maximum results, keep soil under tree cultivated shallowly. Water in dry weather.

At planting time, cut tree back to about one-half its original height to compensate for damage done to the roots. Make the cut 1″ above a bud. Over the next 3–4 years prune branches to develop a strong, well shaped top. Don't remove all the lowest branches at once, but in the end the lowest branch should be 4′–5′ above the ground. The next should be 12″–18″ above this and a third of the way around the tree; the next should be 12″–18″ above this and a third of the way around the tree, and so on.

Spray with malathion to control aphids. Spray with Bordeaux mixture 4–5 times from the time leaves are half formed until August 1 to control scab. Pick up and burn twigs found under trees to get rid of twig girdlers.

PENTAS LANCEOLATA Egyptian star cluster. Three-foot shrub with star-shaped flowers in many colors. Grows in warmest climates; also a house plant.

Outdoors. Propagate by semi-hardwood cuttings in early spring. Plant in spring or fall in average soil in partial shade. Water in dry weather. Fertilize 2–3 times a year. Protect from frost if you want almost continuous flowering.

Indoors. Pot in new general-purpose potting soil every February. Grow in a south window. Water when soil feels dry and spray foliage occasionally with water. Fertilize every 2–4 weeks. From November through January, reduce water and stop fertilizing. Move outdoors in summer.

PENSTEMON Beard tongue. Large genus of perennials and shrubs to 6′ with handsome clusters of flowers in many colors. Grows best in mild climates.

Propagate by division in early spring or fall or by seeds sown in the spring (some species will flower in the same year). Grow in average soil to which extra sand has been added to improve the drainage. Fertilize in early spring and before flowering. Keep well watered during summer. Cover with a light winter mulch. Penstemons grow best in partial shade but do well also in sun.

PEONY *Paeonia.* The familiar peony is a perennial to 3′ with magnificent, fragrant, white, pink or red flowers in spring. It grows in all but the warmest climates. The tree peony is a shrub to 6′ with similar flowers. It grows best in temperate climates.

Perennial. Propagate in August by cutting the large roots into sections with 3–5 eyes. Plant 1″ deep and at least 3′ apart in airy, sunny spot. The soil must be well-drained, humusy and enriched in early spring and again six weeks later with fertilizer. Cultivate and water regularly. Keep wilted or diseased stems and foliage cut out and burn all trash. Also to prevent disease, spray every two weeks in the spring and summer with zineb. Peonies do not have to be moved or divided unless they are obviously not doing as well as they should.

Tree peony. Buy plants in the fall and plant in deeply dug soil with the graft 4″ below ground level. Protect for the first winter with a mulch that is applied after soil freezes. Tree peonies should not be planted in windy locations. They do best in an eastern exposure where they do not get late afternoon sun. Fertilize in early spring. Water regularly. To control scale, spray in late winter with miscible oil and during spring and summer with malathion. Remove dead flowers. Don't cut back woody branches.

PEPEROMIA One species called watermelon begonia. Plant to 2′ with silver or cream-marked, heart-shaped leaves. Grows indoors.

Propagate by stem or leaf cuttings. Pot in two parts loam, two parts humus and one part sand. Grow in a warm north window. Stand pots on wet pebbles. Keep soil moist. Fertilize every 2–3 months.

PEPINO *Solanum muricatum,* melon pear. Shrubby perennials to 3′ with blue flowers and edible, egg-shaped, yellow fruits with aromatic flavor. Grown as an annual in cool climates.

Grow like tomato (which see).

PEPPER *Capsicum frutescens.* Handsome vegetable to 3′ with large, sweet fruits or small, hot fruits. Grows almost everywhere. Hot peppers are also grown indoors.

Outdoors. Grow like tomato (which see) but fertilize only twice, and then sparingly. Plants are spaced about 18″ apart in rows 30″ apart.

Indoors. Sow seeds outdoors in June or July. Pot in individual 4″ pots filled with general-purpose potting soil. Keep watered and in sun. Bring in before frost and grow in cool, sunny window. Water regularly and spray foliage with water occasionally. Fertilize lightly every six weeks.

PEPPER TREE *Schinus.* Evergreen trees to 40′ with finely cut foliage and red or rosy berries in fall and winter. Grows in warmest climates.

Plant in spring or fall in sun in average soil. Male and female specimens are needed if you want berries. Water deeply in dry weather. Fertilize every 2–3 years in the spring. Prune in winter. Spray the California pepper tree with malathion to control scale.

PERESKIA One species called Barbados gooseberry, lemon vine. Shrubby and vine-like cacti to 20′ with woody stems, spines and clusters of flowers in various colors. Grows outdoors in hot, dry climates. Also a house plant.

For how to grow, see cactus, desert. *Pereskias* must be planted in large pots and kept at 50° or above at all times. Water heavily when making growth. Root cuttings immediately, without drying them off.

PERISTROPHE ANGUSTIFOLIA AUREO-VARIEGATA Plant to about 2′ with green and yellow leaves and pink flowers. Grows indoors.

Propagate by stem cuttings. Pot in general-purpose potting soil. Grow in an east, west or south window. Stand pots on wet pebbles. Keep soil moist. Fertilize every 6–8 weeks in fall, winter and spring.

PERIWINKLE *Vinca.* Two species are evergreen groundcovers to 8″ with blue flowers. The better of the two is *V. minor* (also called myrtle) which grows almost everywhere. Madagascar periwinkle (*V. rosea*) is an erect, tender perennial with pink or white flowers. It is usually grown as an annual.

Groundcover. Propagate by layering. Plant in shade or sun in any soil. Fertilize lightly and water while it is getting established; thereafter, water only in long dry spells and fertilize only when it needs a shot in the arm. Plants can be moved at any time.

Perennial. Sow seeds indoors about 10 weeks before last frost. Shift into individual small pots in a sunny window. Don't water too much. When weather is reliably warm, plant outdoors in sun or light shade in good, humusy soil. Fertilize lightly. Water regularly. In warm climates, the plants will survive the winter.

PERNETTYA MUCRONATA Prickly heath. Evergreen shrub to 3′ with red, lilac or white berries that hang on through the winter. Grows in warm climates.

Propagate by stem cuttings in summer. Plant in spring or fall in sun. The soil should be moist, well-drained, of average quality but with a little extra peatmoss. Water in dry weather. Fertilize every 2–3 years. Prune in winter to control shape. Keep roots cut back fairly close to the plant, otherwise they may invade the whole garden.

PERSIMMON *Diospyros.* Deciduous trees to 50′ with handsome foliage and edible, orange or yellow fruits. The common persimmon grows in the south and as far north as the Great Lakes. The Japanese persimmon, which is the best type fruit, grows only in warm climates.

Plant young male and temale specimens 20′ apart in early spring. Grow in full sun in average soil which must be well drained. Keep well watered during spring, summer and early fall. Cultivate regularly. Fertilize in the fall with well-rotted or dried manure. Prune in later winter just enough to keep tree open and well shaped.

PETUNIA Popular annual to 2′ which grows anywhere. Newest varieties are mostly compact and bushy. Bedding types spread fairly wide, can be used in hanging baskets.

Sow seeds indoors 8–10 weeks before last frost. Since the seed is very fine, simply scatter it on the surface of the soil and press in lightly. Transplant to larger flats or small individual pots once before setting out in the garden when the weather is reliably warm. Pinch tips once or twice to promote bushiness.

Petunias bloom best in full sun but will stand some shade. The soil should be well drained, of average quality. Fertilize as soon as newly set out plants are established and again about a month later. Water in dry weather. Keep blooms picked off and cut out dead stems. Burn diseased plants. Spray with malathion if necessary to get rid of aphids.

Petunias are easily grown in pots or window boxes. Treat as above, and never neglect watering.

PHAEDRANTHUS BUCCINATORIUS Blood-red trumpet vine, Mexican trumpet vine. Evergreen vine to 70′ with large, red, trumpet-shaped flowers from spring through fall. Grows in warmest climates.

Propagate by stem cuttings. Plant in spring or fall in rich soil with plenty of humus. Keep moist. Fertilize several times while making growth. Requires sun and good air circulation. Train to a sturdy support. Thin dense growth in center of plant in early spring or late fall.

PHILLYREA Sometimes spelled *filaria.* Evergreen shrubs to 8′ with longish, glossy leaves, small white flowers in spring and small black fruits. Grows in warm climates.

Plant male and female specimens if you want fruits. Propagate by stem cuttings in summer. Plant in fall or spring in sun in average soil. Water in dry weather. Fertilize every 2–3 years in spring.

PHILODENDRON Reliable, handsome foliage plants, usually vines to 20′ but a few shrubby. Mainly a house plant. However, several shrubby types grow outdoors in warmest climates.

Indoors. Propagate by stem cuttings in water (large-leaved varieties should be propagated in a sand-peatmoss mixture, however). Pot in general-purpose potting soil. Grow in any warm, bright window; philodendron doesn't need sun, but it definitely needs more light than it gets in an inside corner of a living room. Water when soil feels dry. Spray foliage with water at least once a week. Fertilize lightly every month. Repot in fresh soil every couple of years. To promote branching, pinch stem ends occasionally.

Outdoors. Plant in average, well-drained soil containing a lot of humus. Grow in partial shade. Water regularly and feed monthly.

PHLOX *Phlox* may be divided roughly into three categories: the showy, erect perennials to 4′, of which the outstanding species is *P. paniculata* (hardy garden *phlox*); the low-growing perennials, notably *P. subulata*, also known as moss pink; and the annuals. One or more species of *phlox* grows almost everywhere in the country.

Tall perennials. Divide plants every three years in spring or fall, but use only the young outer growth. Plant in good, deeply dug, well-drained soil to which humus has been added. Fertilize in early spring and twice more at monthly intervals. Water and cultivate regularly. Pick off dead flowers. Spray with a combination insecticide-fungicide. Mulch in winter. These *phlox* grow well in partial shade but do better in sun.

Creeping phlox. Divide in spring or fall or make tip cuttings in midsummer. Grow in sun in average, well-drained soil. Fertilize in early spring and again after flowering. Water in dry weather. These phlox are very amenable but appreciate a little better care than they usually receive.

Annual phlox. Sow seeds indoors about eight weeks before last frost, or sow outdoors where plants are to grow when weather is warm. Grow in sun in average soil. Water regularly. Fertilize when young plants are set out or thinned. Pinch back some to encourage branching. Keep dead flowers picked off.

PHOTINIA Deciduous and evergreen shrubs to 30′ with white flowers and long-lasting red fruits. The evergreen species grow in warm climates; the deciduous types are hardy further north.

Propagate by cuttings or by layering. Plant in spring or fall in a sunny location in a light, well-drained, average soil. Water in dry spells. Fertilize every 2–3 years. Pinch out tips of new stems. Prune in winter to control shape.

PHYLLOSTACHYS Genus of bamboos. Grows in warmest climates.

For how to grow, see bamboo.

PICK-A-BACK PLANT *Tolmiea menziesi.* Plant to 15″ with hairy leaves on which little new plants grow. A house plant.

To propagate, cut off a leaf with a little growing plant, and pin it down on damp sand. Pot rooted plant in general-purpose potting soil. Grow in an east or west window at a temperature of 65°. Keep soil moist. Fertilize every 3 months.

PICKEREL WEED *Pontederia cordata,* pickerel rush, pond weed. Perennial to 2′ with arrow-shaped leaves and blue flower spikes. Grows in marshes in the east.

Divide in early spring and plant in partial shade in average soil under 3″–4″ of water.

PIERIS JAPONICA Japanese Andromeda. Handsome evergreen shrub to 9′ in panicles. Grows best in warm and mild climates.

Propagate by layering. Plant in spring or fall in light shade. The soil should be of average quality and acid. Add peatmoss for moisture retention and sand for good drainage. Water in dry spells. Fertilize annually in early spring with a 10-8-8 plant food. Spray several times in spring with malathion to control lacebugs, or apply a systemic poison.

PILEA Different species called aluminum plant, artillery plant, friendship plant, panamiga. Foliage plants to 15″ of widely varying character. Grows indoors.

Propagate by stem cuttings. Pot in general-purpose potting soil. Grow in an east or west window. Keep evenly moist. Fertilize occasionally. Stand pots on wet pebbles.

PINCUSHION CACTUS *Mammillaria.* Different species called thimble cactus, golden stars, powder puff cactus, feather cactus, silken pincushion, old lady cactus, mother of hundreds, bird's nest cactus, snowball pincushion, owl's eyes. Desert cacti to 1′ with lovely flowers in various colors. Grows in hot, dry climates; also indoors.

For how to grow, see cactus, desert. House plants should be repotted in fresh soil every year for the first five years.

PINE *Pinus.* Most of these long-needled conifers are tall, shapely trees to as much as 180′, but mugo pine is a useful shrub which can be held to only a few feet. Different species grow in different regions.

Plant in spring or early fall. Like most conifers, pines grow best on a north slope, where they are protected from scorching winter sun and hot westerly winds; but they do well in any sunny location that is protected to some extent from dry winds. The soil should be of average quality and well drained. Don't expose roots directly to the air when transplanting. Water deeply in drought. Fertilize young plants in spring for a couple of years if you want them to get off to an exceptionally fast start. Remove dead branches. Little other pruning is required except on mugo pines. To control the growth of these, prune in the spring just before the needles on the candles (new tip growths) open out. Trim the candles back to about half their length.

Pines are subject to attack by several diseases and insects. Treat as follows as necessary: Spray with malathion twice at ten-day intervals in spring to control needle scale. Spray with oil emulsion in late winter to control bark aphids. Eliminate currants or gooseberries in the neighborhood if blister rust attacks white pines.

PINEAPPLE *Ananas.* A bromeliad with sword-shaped leaves in a rosette and the familiar fruit on a stalk up to 4′ tall. Grows in warmest Florida; also indoors.

Outdoors. Propagate by suckers. Or cut the rosette of leaves from the top of a fruit, being sure to take a thin slice of the fruit. Allow to dry for 24 hours, then sprinkle with a transplanting hormone and set in sand or soil just deep enough so that it will stand upright. Keep barely moist and in deep shade until roots form. Then move into half sun in acid, rich, well-drained, sandy soil. Space plants 2′ apart. Water in dry weather but not too much to encourage rot. Fertilize with a 6-4-8 plant food worked lightly into the soil close to the base

of the plant. Make applications every month while plant is making growth. After plant has fruited once it is through. To prevent nematode attack, fumigate soil before pineapple is planted. Pick off mealybugs with a cotton swab dipped in alcohol.

Indoors. Grow from a rosette of leaves as above. Keep in a warm, sunny window after the roots have developed. Water when soil feels dry. Feed every three weeks with liquid plant food. Move outdoors into partial shade in summer.

PINEAPPLE GUAVA *Feijoa sellowiana.* Wide-spreading, evergreen shrub to 12′ with purple and red flowers in spring and oblong, grey-green, edible fruits in fall. Subtropical.

Plant in spring or fall in a sunny location in well-drained loam containing considerable humus. As a rule, you need to plant two different varieties to assure pollination; however, Coolidge can be grown alone. Water in dry seasons. Fertilize lightly with bone meal or general-purpose plant food with a low nitrogen content once a year. Work additional humus into the soil every couple of years. Prune in late winter.

PINEAPPLE LILY *Eucomis comosa.* Bulbous plant to 2′ with greenish-white summer flowers. Grows almost everywhere; also indoors.

Outdoors. When weather is reliably warm, plant bulbs 1″ deep and 1′ apart in well-drained, average soil in a sunny spot. Fertilize once. Water in dry weather. Dig up and store bulbs in peatmoss in a cool, dry place in winter. In the deep south, bulbs may be left in the ground.

Indoors. Grow like freesia (which see).

PIPSISSEWA *Chimaphila.* Evergreen wildflower to about 1′ with leathery leaves and white flower clusters. Grows almost everywhere.

Buy plants. Set out in spring in light shade in well-drained, somewhat acid, humusy soil. Water in dry weather.

PISTACHIO *Pistacia vera.* Evergreen tree to 30′ with a dry fruit containing a green seed. Grows mainly in hot, dry parts of California.

Plant one-year-old specimens in

spring in sun in deep, rich soil. Space trees 30′ apart. One male tree is needed with every 5–6 female trees. Water well during the first year; thereafter, in dry weather only. Fertilize in spring. Prune in winter to maintain shape.

PITCHER PLANT *Sarracenia,* trumpets, trumpet leaf, green regent. Perennials to 3′ with pitcher- or trumpet-shaped leaves, often colored and containing a nectar which attracts insects which the plant eats. Grows mainly east of the Mississippi; also indoors.

Outdoors. Divide in early spring and plant in partial shade in moist, acid, very humusy soil.

Indoors. Pot in sphagnum moss or peatmoss mixed with sand. Keep moist all the time (don't use alkaline water). Grow in an east or west window. Move outdoors in summer.

PITTOSPORUM TOBIRA Australian laurel. Evergreen shrub to 10′ with leathery leaves and white flower clusters. Grows outdoors in warmest climates; also indoors.

Outdoors. Propagate by semi-hardwood cuttings. Grow in sun or partial shade in average soil. If grown for a hedge, space 30″ apart. Water in dry weather. Fertilize every 2–3 years. Prune any time to control growth and shape.

Indoors. Pot in general-purpose potting soil and grow in a south window at no more than 55°. Water when soil feels dry. Fertilize 3–4 times during the year.

PLUM *Prunus.* Deciduous trees and shrubs with white, pink or red flowers and red, purple or green fruits. Grows in mild and warm climates.

Plant two varieties of European plum or two varieties of Japanese plum in spring in full sun on a slope—preferably a northern slope. The soil should be deeply dug, well-drained, and of average quality. Space trees 20′–24′ apart and set them a little deeper in the ground than they formerly grew. Water in dry spells until the end of August; thereafter, don't water. Fertilize in early spring at the rate of 1 lb. of general-purpose plant food per year of tree age up to a maximum of 10 lb. Prune in late winter just enough to trim out dead wood and thin out extraneous branches. Japanese plums require more pruning than European types.

To control insects and disease, spray as for cherries (which see).

PLUMBAGO Leadwort. Sprawling, evergreen plants to 8′ that can be trained as climbers, with red, white or blue flowers in clusters in summer. Grows outdoors in warmest climates; also a house plant.

Outdoors. Propagate from seeds started in February or by cuttings of side shoots in February or August. Plant in spring in sandy soil in the sun. Don't water too much. Fertilize in early spring. Tie branches to a strong support. Cut out suckers and prune hard in early spring.

Indoors. Pot in general-purpose potting soil. Water when soil feels dry. Fertilize lightly every six weeks in spring and early summer. Keep in a sunny window and move outdoors in summer. In the fall, cut back severely and repot. Then keep in a cool window and water sparingly until early spring, when heat and watering should be increased.

PLUM YEW *Cephalotaxus* Handsome evergreen trees or shrubs to 30′ with needle-like leaves and egg-shaped fruits. Grows best in mild sea-coast climates.

Plant in spring or early fall in a sunny or partially shaded location. Grow in average, well-drained soil. Water in dry weather. Fertilize every 3–5 years. Prune in spring.

PODOCARPUS Different species called yew pine and fern pine. Graceful, irregular, evergreen trees or shrubs to 20′ with narrow, dense leaves. Grows in warmest climates; also indoors.

Outdoors. Plant in spring or fall in sun or partial shade in well-drained loam soil. Water regularly while young; thereafter, in dry weather. Fertilize in spring every 3–5 years. Prune in winter. Young plants need protection from wind.

Indoors. Propagate by stem cuttings rooted in moist sand. Plant in pots or tubs of general-purpose potting soil. Grow in a cool, south window Water when soil feels dry and fertilize a couple of times a year Spray foliage with water every week. Shape foliage with shears if plant threatens to become leggy.

POINCIANA One species called bird-of-paradise bush; another, dwarf poinciana or Barbados pride. Deciduous shrubs to 10′ with yellow and red flowers. Grows in warmest climates. See also royal poinciana.

Sow seeds in spring in pots or flats of average soil in light shade. The seeds should first be soaked in warm water for several hours. Transplant into garden in spring or fall when plants are about 1′ tall. Poinciana needs full sun, an average soil. Once established, watering is required only in dry spells. Fertilize every spring. Prune in spring.

POINSETTIA *Euphorbia pulcherrima.* Ever-popular Christmas plant with showy red or white bracts. Grows indoors; also outdoors in warm climates.

Indoors. Keep Christmas plants in an east or west window at no more than 65°. Remove wrappings and set pot on wet pebbles. Keep soil moist but not soaking. When bracts fade, gradually withhold water and stop watering entirely when foliage yellows. Then store plant in a cool (under 60°), out-of the-way place until May. Then cut stems to 6″ length, repot in new soil, water and set in a south window at 65°. When new growth has made a vigorous start, take tip cuttings and root them in sand. Then pot up in small pots of general-purpose potting soil, and when weather is warm, plunge the pots outdoors in a sunny spot. Water regularly. Fertilize every 3–4 weeks. As plants grow, move into larger pots. Bring indoors before frost and set in a cool, south window. Continue watering and, about November, increase feeding frequency to every two weeks. From October 1 on, expose plants only to natural light—not to electric light after sunset.

Outdoors. Propagate by stem cuttings in early spring. Plant these when they root in a sunny, sheltered spot in average soil. Water regularly and fertilize every six weeks. To promote bushier growth, pinch out stem ends when plant is young. In September, cut out all but about five stems and reduce these one-third. After flowering ends, if you intend to continue with the old plants instead of starting new ones from cutting, cut stems to within 6″ of the ground. Water sparingly until new growth starts.

POLEMONIUM *P. caeruleum* called Jacob's ladder or charity; *P. reptans* called bluebell or American abscess root. Perennials to 3′ with clusters of blue or white flowers. Grows in temperate climates.

Divide *P. caeruleum* in the fall; *P. reptans* in spring. Plant in sun or partial shade in average soil. Water in dry weather. Feed in spring.

POLYPODIUM Creeping ferns to about 10″ with narrow, open fronds. Polypody or wall fern grows widely throughout the country. Resurrection fern grows on trees in the south. Hare's foot fern or golden polypody is a house plant.

For how to grow, see fern. To propagate the resurrection fern, cut rooted pieces from a plant and pin into the folds of bark on a tree. Water well until it is established. This plant is unusual in that it curls up and looks dead in dry weather, but quickly revives when it rains.

POLYSTICHUM Popular species known as Christmas fern, mountain holly fern and giant holly fern. Plants usually are evergreen, to 3′. Grows in temperate climates.

For how to grow, see fern. These ferns do best in considerable shade and in cool, moist, humusy soil. Divide and plant in early spring.

POMEGRANATE *Punica granatum.* Deciduous shrub to 20′ with orange-red flowers and large, edible, seedy, red fruits. A dwarf variety, almost evergreen in mild winters, is grown only for ornament. Grows in warmest climates; also indoors.

Outdoors. Plant in spring or fall in full sun in average soil. Water in dry weather. Fertilize every spring. Remove suckers that develop at the base; they are easily propagated.

Indoors. Pot in a tub of general-purpose potting soil. Keep moist. Grow in a south window in a cool (about 60°) room. Start fertilizing lightly in the spring and make applications every six weeks through the summer. Move outdoors in summer.

POND LILY *Nuphar*, spatterdock, cow lily. A small, yellow water lily growing east of the Rockies.

Grow like water lily (which see).

PONGAMIA PINNATA Handsome tree to 40′ with almost evergreen foliage and fragrant, white flowers. Grows in warmest climates.

Plant in spring or fall in average soil in sun. Water in long dry spells. Fertilize every 3–5 years. The tree is very resistant to wind and salt spray.

POPCORN *Zea mays everta.* A variety of corn growing to about 6′. Grows almost everywhere.

Plant in the spring about the time of last frost and grow like sweet corn (see corn). Allow ears to mature fully; then cut stalks and shock them like field corn to dry completely. Then remove ears, shuck and store in a dry place. Kernels can also be removed from ears.

POPLAR *Populus.* Different species called aspen, cottonwood, tacamahac, balm of gilead. Deciduous trees to 90′ with soft wood and oval leaves that turn bright yellow in fall. Grows almost everywhere.

Plant in spring or fall in a sunny location in average soil. Don't plant close to a sanitary drainage field, because the roots will soon invade the pipes. Water in dry spells, especially the cottonwood. Keep dead and broken wood pruned out. Spray in late winter with miscible oil to control scale. Use malathion to control poplar-leaf beetle. Don't expect poplars to live long.

POPPY *Papaver*. The true poppy family includes the handsome perennial, oriental poppy, which grows to 4′ in all but the warmest climates; miscellaneous perennials, most notably the Iceland poppy, which grow best in cool climates; and several annuals, particularly the Shirley poppy, which grow anywhere.

Oriental poppy. Propagate by the long roots in late summer. Cut these into 4″ lengths and plant right-end up 3″ deep and 15″ apart in a sunny spot. Grow in average, well-drained soil to which humus has been added. Fertilize in early spring. Water regularly during spring and early summer, but do not water after flowering. Mulch soil the first winter after roots have been planted. Stake tall plants which are exposed to wind.

Other perennial types. Sow seeds where plants are to grow in the early spring. Grow in sun in average, well-drained soil. Fertilize when plants are thinned to 9″. Water regularly. Plants are not sure to survive the winter, although they may self-sow and gradually revert to type. It may be advisable, therefore, to treat Iceland and similar perennial poppies as annuals.

Annual poppies. Mix seed with sand and sow where plants are to grow in very early spring or the previous autumn (except in coldest climates). Don't cover the seeds with soil but water well. Grow in sun in average, well-drained soil. Fertilize lightly in the spring when plants are thinned to 9″–12″. Water in dry weather. Keep dead flowers picked off.

POPPY MALLOW *Callirhoe*. Perennials to 2′ with showy purple flowers in spring and summer. Grows mainly in midwest and south.

Divide in spring or fall and plant in sun in average, well-drained soil. Water in dry weather. Fertilize in spring.

PORTULACA Rose moss, sun plant. Annuals to 8″ covered with bright flowers in many colors. Grows anywhere.

Sow seeds in late fall or early spring in a sunny location in light, well-drained soil. Thin to 6″ apart and forget them.

POTATO *Solanum tuberosum*. Vegetable to 30″. Grows almost everywhere but does best in cool weather. Use variety recommended for your area.

Potatoes grow well in many different soils. For best results, these should be well-drained and aerated, not too stony, humusy, slightly acid, deeply dug and enriched with 5-10-10 fertilizer which is applied several weeks before the "seeds" are planted.

Propagate potatoes from tubers which are certified to be free from disease. Cut tubers into several pieces each with 1–2 eyes and as much flesh as possible, and soak for 1½ hours in a solution of corrosive sublimate. Plant the pieces 5″ deep and 1′ apart in rows 30″–36″ apart. Early potatoes are planted about two weeks before date of last spring frost; late varieties go in in time to mature just before the first autumn frost.

Cultivate regularly and carefully —especially while plants are young —to keep out all weeds. At final cultivation, mound up soil slightly around the plants so that the tubers will be well covered. Irrigate deeply in dry weather. To control insects, spray or dust with DDT or methoxychlor several times during the growing season. Dig early varieties at any time. Dig late varieties, which are to be stored, when the tops wither. At this time, let the tubers lie in the sun until dry, then store them in a cool, dark place.

POTATO VINE *Solanum jasminoides.* Slender-stemmed vine to 30′ with dainty, blue-white, summer flowers. Vine is evergreen or deciduous, depending on the severity of the winter. Grows in warm climates.

Propagate by stem cuttings. Plant in average soil in sun or partial shade. Protect from winds. Keep watered. Fertilize in spring. Tie branches to a trellis. Cut back hard in early spring.

POTENTILLA Cinquefoil, buttercup shrub. Perennials and shrubs to 4′ with hairy leaves and clusters of yellow, white or red flowers. Grows in temperate climates.

Perennial. Divide every 3–4 years in spring or fall and plant in sun in average soil. Water in dry weather. Fertilize in spring.

Shrub. Propagate by stem cuttings of half-ripe wood in early fall. Plant in spring or fall in sun or partial shade. Prune in early spring. Otherwise treat as above.

POTHOS *Scindapsus,* ivy arum, hunter's robe. Evergreen, climbing or trailing plants to 40′ with variegated leaves. A house plant.

Pot in general-purpose potting soil with extra humus. Grow in a warm east or west window. Water when soil feels dry. Fertilize 3–4 times a year. Pinch stem ends occasionally to encourage branching and leaf development at the base of the plant. Provide rough wood for the plant to climb on.

PRIMROSE *Primula.* Perennials to 2′ but usually smaller, covered with gay flowers in many colors in spring. Grows best in cool, moist climates. Several tender species—*P. obconica, P. malacoides erikssoni* and *P. sinensis*—are good house plants.

Outdoors. Divide plants every 3–4 years after flowering. Or sow seeds as soon as they are harvested in the fall in a coldframe and do not move the plants until spring. Or sow seeds indoors in the spring about eight weeks before last frost. In this case, they should be frozen and thawed repeatedly in an ice tray for half a week.

Plant primroses in spring or fall in partial shade in a well-drained, rich soil which contains considerable humus. Set the crowns a little above the soil level. Keep moist. Fertilize in early spring and once more about six weeks later. Cultivate carefully to avoid disturbing the roots. Mulch in winter. Spray regularly with a combination insecticide-fungicide. Be sure that the spray strikes the underside of the leaves.

Indoors. Sow seeds in January and move plants into successively larger pots of general-purpose potting soil. Grow in a very cool east window. Keep soil moist. Fertilize every 2–3 months. In summer, move outdoors into a cool, partially shaded place.

PRINCESS FLOWER *Tibouchina,* spiderflower. Evergreen shrubs or trees to 20′ with large purple flowers in spring, summer and fall. Subtropical.

Propagate by semi-hardwood cuttings in spring. Plant in spring or fall in partial shade (*T. granulosa* also grows in sun) in average, well-drained, somewhat acid soil. Water in dry weather. Fertilize in spring. Prune hard in late winter and occasionally during warm weather after blooming periods to prevent legginess.

PRINCESS PALM *Dictyosperma album.* A graceful, single-stemmed palm tree to 40′ with upward-reaching, open fronds and dense orange flowers. Grows in subtropics.

For how to grow, see palm.

PRIVET *Ligustrum.* Fast-growing, evergreen and deciduous shrubs to 20′, most commonly used for hedges. Grows in all but coldest climates.

Privets for hedges are propagated by young stem cuttings rooted in sand in June. Set out plants in spring or early fall. Privets are easygoing, doing well in sun or partial shade, in city and industrial atmospheres and in average soil. For the most luxuriant hedge, however, handle as follows:

Start with two- or three-year-old plants. Dig amply large holes and fill with topsoil mixed with a little peatmoss and some fertilizer. Set the plants 2″–3″ deeper than they formerly grew. Space them 9″–12″ apart in the row. After planting, cut branches back about a third. Water heavily several times for the next few weeks.

During the first summer, trim plants about three times (but not after Labor Day, except in warm climates). Let the branches grow about 1′, and then cut back one half. In this way, plants are forced to bush out. The following year, when the hedge begins to develop its proper shape, clip as often as you want for the sake of neatness, but remember that the hedge must be wider at the bottom than at the top, otherwise it will become naked at the bottom.

Once a privet hedge has become established, it needs little attention, though it responds well to a little fertilizer every 1–2 years. Water in long dry spells.

PSEUDERANTHEMUM Shrub to 3′ with colored foliage and white or purple flowers. A house plant.

Propagate in spring by stem cuttings. Pot in general-purpose potting soil. Grow in a warm east or west window. Keep soil moist and spray foliage with water every week. Fertilize 2–3 times in fall and winter.

PSEUDOSASA Genus of bamboos. Grows in warmest climates.

For how to grow, see bamboo.

PTEROSTYRAX HISPIDA Epaulette tree. Deciduous tree to 45′ with large leaves and fragrant, white flower panicles in spring. Grows in mild and warm climates.

Plant in spring or fall in sun in

somewhat sandy soil which retains moisture. Water in dry weather. Fertilize every 3–4 years.

PTYCHOSPERMA One species called solitaire palm. Tree to 30′ with single or clustered stems and feather-like fronds with blunt leaves. Subtropical.

For how to grow, see palm. Fertilize four times a year.

PULMONARIA *Lungwort.* One species called Bethlehem sage. Perennials to 18″ with funnel-shaped flowers in white, pink or purple. Grows almost everywhere.

Divide in spring or fall and plant in partial shade in average soil. Water in dry spells. Fertilize in spring.

PUMPKIN *Cucurbita.* Large sprawling vine with huge orange fruits. Grows almost everywhere that weather stays warm for about four months.

The soil for pumpkins must be light, well-drained and rich. Spade in humus and fertilizer. Sow seeds in full sun after all danger of frost is past. Sow 8-10 seeds 1½″ deep in a 1′ circle (hill) and thin later to two plants. The hills should be at least 6′ apart in all directions. Water regularly and deeply. Pull weeds. Sidedress with fertilizer when plants are about six weeks old. Until flowers begin to bloom, dust frequently with rotenone to discourage insects. Harvest before frost.

PUSCHKINIA Striped squill. Bulbous plant to 1′ with dainty, grey-blue spring flowers. Grows best in temperate climates.

Plant bulbs 3″ deep and 3″ apart in partial shade in the fall. The soil should be well drained, of average quality. Fertilize every couple of years.

PUSSY WILLOW *Salix discolor.* Deciduous shrub to 18′, prized for its catkins in early spring. Grows almost everywhere.

Propagate by stem cuttings in the spring. Plant in average, moist soil in a sunny spot in spring or fall. Water in dry weather. Cut back hard in the spring after catkins form to promote more catkins the next year.

PYRACANTHA Firethorn. Evergreen shrub to 20′, noted for its large clusters of orange-red berries in the fall. Grows in all but coldest climates.

Propagate by layering. Plant in spring or fall in average, well-drained soil. *Pyracantha* prefers full sun but does well in partial shade. Water in dry weather. Fertilize in the spring every 1–2 years. Prune in late winter to control growth. Plants can easily be trained as vines or espaliers.

PYRETHRUM *Chrysanthemum coccineum*, painted daisy. Perennial to 2′ with showy flowers in various colors in late spring and early summer. Grows in temperate climates.

Divide plants every three years in early spring or July. Discard woody growth. Plant in sun in well-drained, average soil improved with humus. Water regularly. Fertilize in early spring and after flowering. Cut flower stems to the ground after flowering.

QUAMOCLIT Different species called cypress vine, star glory, Spanish flag, cardinal climber. Mostly annual vines to 20′ with fern-like foliage and scarlet summer flowers. Grows in warm climates.

Soak seeds overnight in water before sowing. For earliest bloom, start seeds indoors; otherwise, sow them in a sunny spot after danger of frost is past. Grow in average soil which is kept moist. Fertilize when young plants are established and once or twice more. Train on string or wire.

QUEEN'S WREATH *Petrea volubilis*, purple wreath, sandpaper vine. Evergreen vine to 20′ with long clusters of blue flowers in spring and summer. Subtropical.

Propagate by stem cuttings. Plant in rich soil with lots of humus. Keep evenly moist and fertilize 2–3 times during growing season. Grow in sun or partial shade and protect from wind. Stems twine around wire or a wood lattice. Prune lightly after flowering.

QUINCE *Cydonia oblonga*. Deciduous tree to 20′ with hard, yellow fruits like apples—not very edible but used in jelly and preserves. Grows best in Middle Atlantic states and Ohio.

Plant in spring in full sun in deep, well-drained soil of above-average quality. Space trees 20′ apart. Water in dry weather. Fertilize in spring at the rate of 1 lb. per year of age up to a maximum of 15 lb. Prune in late winter to thin out branches (maintain a spreading, open head). Also cut back branches slightly to stimulate the new growth on which fruits are borne. To control pests, see apple.

RADISH *Raphanus*. Tasty root vegetable with 5″ tops (except winter radishes grow to 2′). Grows anywhere.

Most radishes grow so fast that they are not particular about soil, but they do best when it is well pulverized, deeply dug, cleared of stones and enriched with a little fertilizer. Make succession sowings from the time that the soil can first be worked until early summer; then start all over again in late August. Sow seeds in rows 1′ apart and thin plants to 1″. Water in dry weather.

RAIN LILY *Cooperia*, evening star, fairy lily. Bulbous plants to 1′ with fragrant, pinkish, night-blooming flowers after rain. Native to the southwest but grows almost everywhere.

Plant in spring 2″ deep and 6″ apart in average, well-drained, fairly gritty soil. Grow in sun. Dig up in the fall and store bulbs in peatmoss in a cool, dry place. In warm climates the bulbs may be left in the ground.

RANGOON CREEPER *Quisqualis indica*. Rangy deciduous vine to 25′ with summer flowers which turn from white to pink. Subtropical.

Propagate by seeds or stem cuttings and plant in spring or fall in moist soil improved with humus. Grow in sun out of the wind. Fertilize in spring. Tie stems to a strong support. Prune hard right after flowering.

RANUNCULUS ASIATICUS Florist's ranunculus, Persian buttercup. Tuberous plants to 15″ with flowers like double, yellow buttercups. Grows best in warm climates. Can be grown indoors if a sunny room with a 40°–50° temperature is available.

Outdoors. Soak tubers in water

for two hours before planting in late fall. Plant in good, well-drained, humusy soil 1″–2″ deep, 6″ apart and with the claws down. Water sparingly until leaves appear, then increase supply a little. Fertilize in spring. After flowering, tubers can be dug up and stored in a dry place but it is better to discard them and start with new.

RAPHIOLEPIS Indian hawthorn, yeddo hawthorn. Evergreen shrubs to 10′ with thick, dark green leaves and white, pink or red flowers in spring. Grows in warmest climates.

Propagate by semi-hardwood stem cuttings in summer. Plant in spring or fall in sun in average soil. Water in dry weather. Fertilize every 2–3 years in late winter. Prune after flowering to control shape. Some specimens are spreading, some upright, some irregular.

RASPBERRY, BLACK *Rubus*, black-cap. Bramble fruit to about 8′ bearing black berries. Plants are weeping and produce new plants when the tips touch the ground. Grows in mild and cold climates but not as hardy as red raspberries.

Propagate by layering the tips of the canes. Plant in spring in a sunny spot in average soil containing extra humus and some fertilizer. Space plants 5′ apart in rows 5′ apart. Set them 1″ deeper than they formerly grew and cut off all but 6″ of the top growth. Water well during the summer and mulch to keep down weeds.

In order to keep plants from rooting at the tips when you don't want them to it is necessary to keep them low and bushy. Snip off canes to a height of 18″ whenever they grow much beyond this. This forces branches to develop. These should be cut back each spring to six buds.

In addition, old canes should be removed completely after they have borne fruit. New canes should be removed every spring so that plant has only about five large ones.

Regular annual culture of black raspberries also calls for application of a handful of fertilizer in the spring; regular watering if rain doesn't fall; a summer mulch.

Note. Don't plant black raspberries within ¼ mile of red raspberries since they may be damaged by a disease which is carried by (but does not affect) the reds.

RASPBERRY, PURPLE *Rubus.* A close relative of the black raspberry, but with purplish fruit.

Grow like black raspberry (see above).

RASPBERRY, RED *Rubus.* Bramble fruit to 8′ producing red berries on year-old canes. Plants grow from suckers. Some varieties called "everbearing" because they produce fruit on new canes the first fall and a second crop the next summer. Red raspberries grow best in mild and cold climates.

Propagate by the suckers. Plant in the spring (in milder climates, however, fall planting is permissible). Grow in average, well-drained soil containing extra humus and some fertilizer. Set plants 2″–3″ deeper than they formerly grew, and cut off tops 6″ above the ground. Space plants 4′ apart in rows 5′ apart. Water thoroughly at planting; thereafter, water after a week without rain. Red raspberries demand lots of water. Keep soil free of weeds during the first summer. Applying a mulch of straw or leaves helps.

From the second year on, follow this procedure: In early spring work general-purpose plant food into the soil at the rate of one handful per plant. Cut all dead canes to the ground and burn them; remove all but 5–6 live canes from each plant and shorten the remainder to 24″–30″. Snip off cane ends of ever-bearing varieties. Pull up all suckers whenever they appear during the year. Water well during the growing season and keep soil mulched.

RASPBERRY, YELLOW *Rubus.* A close relative of the red raspberry but with yellow fruit.

Grow like red raspberries (see above).

RAT-TAIL CACTUS *Aporocactus flagelliformis.* Desert cactus with thin, hanging stems 3′ long and red spring flowers. Grows in warmest climates; also indoors.

For how to grow, see cactus, desert. This cactus is a favorite for growing in hanging pots. Water regularly when making growth. Don't expose to frost.

RECHSTEINERIA One species called Brazilian edelweiss. Tuberous plants to 1′ with orange or red flowers and velvety foliage. Grows indoors.

Plant tubers in fall in two parts loam, two parts humus and one part sand. The soil should just cover the tuber. Grow in a south window but provide midday shade in summer. While plant is making growth, fertilize every three weeks and water when soil feels dry. After flowering, stop feeding and reduce water supply gradually. When foliage dies down, store tubers in a cool, dry place until December, when they can be started into growth again.

REDBUD *Cercis.* Deciduous trees to 30′ with purplish-pink flowers in early spring. Grows in mild and warm climates.

Plant young specimens only in the spring. Grow in sun or partial shade in average, well-drained soil. Don't expose plants to strong winds because wood is brittle. Water in dry weather. Feed every 2–3 years in late winter. Spray with malathion to control leaf tiers.

RED HOT POKER *Kniphofia uvaria,* tritoma, torch lily. Perennial to 4′ with spikes of yellow to red tubular flowers. Grows best in warm climates.

Propagate by division in the spring or by seeds sown outdoors when weather is warm. Grow in full sun in well-drained, sandy soil. Fertilize lightly in spring. Water in dry weather. Divide clumps every 3–4 years. Apply a heavy mulch of salt hay in the winter in cold climates, but don't count on the plant's surviving even so.

RED IVY *Hemigraphis colorata.* Trailer a couple of feet long with foliage that is reddish-purple in sun and silver in shade. A house plant.

Propagate by stem cuttings. Pot in general-purpose potting soil. Grow in a north window if you like silvery leaves; an east or west window for purple leaves. Keep moist. Spray foliage with water often. Fertilize 2–3 times a year.

REHMANNIA ELATA Perennial to 3′ with few leaves on the stems and showy, reddish-purple flowers with yellow throats that are spotted red. Grows in warmest climates.

Divide in fall. Plant in average soil in sun or partial shade. Water in dry weather. Fertilize in winter and spring.

RHEKTOPHYLLUM MIRABILE Sometimes identified as *Nephthytis picturata.* Climber to 6′ indoors, with large, handsome, papery leaves with silver markings. A house plant.

Pot in general-purpose potting soil. Grow in an east or west window. Keep soil moist. Spray foliage with water frequently. Fertilize about four times a year. Provide a trellis or bark-covered piece of wood for plant to cling to.

RHIPSALIS Mistletoe cactus, willow cactus. Jungle cactus with thin, hanging, 3′ stems and white flowers. Grows indoors; also outdoors in subtropics.

Grow like orchid cactus (which see). Plant in a hanging container in a mixture of one part loam, one part humus, one part sand and three parts osmunda fiber.

RHODODENDRON Magnificent shrubs, usually evergreen but some deciduous, to 30′, with huge flower clusters in many colors in spring. Grows best in east, southeast and northwest. See also *azalea.*

Propagate by layering notched branches. Plant in spring in a location that is protected from the midday sun and also from cold winter winds. If natural protection from wind is not possible, protect plant in winter with a burlap screen. The soil should be deeply dug, well drained, of superior quality and with a pH of 4.5–5.5. Mix the soil in each planting hole with an equal amount of peatmoss, oak leaf humus or decayed pine needles and with a spadeful of well-rotted or dried manure. If the soil is not naturally acid, mix in aluminum sulfate or powdered sulfur. Water plant well when it is set out, and thereafter water heavily in dry weather and in late fall before the ground freezes. To hold in moisture, maintain soil acidity, keep the soil cool in summer and protect plant roots in winter, maintain a mulch of oak leaves, pine needles or (less good) peatmoss at all times. Do not cultivate.

Given this treatment, the plants should not need much fertilizer, but you can stimulate growth and flowering to some extent by applying a balanced one early every spring. For best bloom, remove faded flowers. If necessary, prune after flowering; however, just pinching out young stem ends should be enough if you start this treatment when the plant is young. Spray with malathion to control insects.

RHODOTYPOS Jet bead, white kerria. Deciduous shrub to 6′ with bright green foliage and large, white flowers in spring. Grows in mild and warm climates.

Propagate by cuttings or self-sown plants. Plant in spring or fall in sun in average soil. Water in dry weather. Fertilize every 2–3 years.

RHOEO DISCOLOR Oyster plant, Moses in the cradle. Plant to about 1′ with large, overlapping leaves and white flowers in boat-shaped bracts. Mainly a house plant, although it grows outdoors in warmest climates.

Pot in general-purpose potting soil. Grow in an east or west window Keep moist. Fertilize lightly 3–4 times annually.

RHUBARB *Rheum.* Long-lived, perennial vegetable grown for its succulent red stalks. Grows best in cool climates.

Rhubarb requires a rich, well-drained soil in a sunny spot where it won't be disturbed. If available, well-rotted manure should be spaded into the soil and used as a winter mulch. Otherwise, mix the soil with a large amount of humus and some general-purpose fertilizer rich in nitrogen.

Propagate rhubarb in early spring by cutting two- or three-year-old roots into pieces, each having 2–3 buds. Plant immediately so that the buds are just below the soil surface. Water well. Space the roots 2′ apart in rows 4′ apart. Don't harvest the stalks until the second spring.

Routine care of established plants consists of the following: Cover with a light mulch of manure or humus in winter. In spring, work this lightly into the soil (take care not to damage roots). Sidedress twice during the spring with nitrogen-rich general-purpose plant food. Detach mature outer stalks from the plant by grasping them at the base and pulling. Do not remove more than half the stalks from any plant in a single season. Keep weeded and watered at all times. Divide the roots every 3–4 years in early spring.

RICE PAPER PLANT *Tetrapanax papyriferum,* Chinese paper plant. Evergreen shrub or tree to 7′ with large leaves at the tops of the stems and white flower clusters in December. Grows in warmest climates.

Propagate by suckers, which must be cut off with as much root as possible. Plant in spring or fall in sun or light shade in average soil. Water well until plant is established, then only in dry spells. Fertilize every 2–3 years. Prune in spring. One serious problem with this shrub is that the roots run wild, so keep them cut back with a spade or grow the plant in a large container.

ROCHEA COCCINEA Shrubby succulent to 2′ with striking, scarlet flower clusters in early summer. Grows indoors.

For how to grow, see cactus, desert. Fertilize every spring and keep well watered. Pinch plants to promote bushiness. Repot in summer into next-size-larger pot. Don't let soil dry out completely for long periods in winter.

ROCK ROSE *Cistus.* Deciduous shrubs to 8′ with flowers, usually purple and yellow or white and yellow, like single roses. Grows in warm climates.

Propagate by cuttings of side shoots in August. Plant in fall or spring in full sun in well-drained, average soil. Unless soil is naturally alkaline, lime it annually. Water in dry weather. Fertilize every 1–2 years. Cut old stems out at the base occasionally in late winter, but not all at once; and don't tamper with young growth. Don't try to move plants once they are established.

ROMNEYA COULTERI Matilija poppy. Perennial to 8′ with white and yellow flowers in spring and summer. Grows in warmer parts of California.

Divide in early winter and plant in sun in average soil. Water deeply every two weeks during flowering.

Fertilize in late winter. Cut almost to the ground in the fall.

RONDELETIA Evergreen shrubs to 15′ with glossy foliage and tubular, pink or orange flowers with yellow throats. Subtropical.

Propagate by half-ripe stem cuttings. Plant in fall in filtered sun in average, sandy soil. Water in dry weather. Fertilize in early winter and again about the time bloom starts in February. Prune after flowering. Spray with malathion to control insects.

ROSE *Rosa.* Roses are available in the following types: hybrid teas, floribundas, grandifloras—the three types most generally planted in rose gardens; hybrid perpetuals—similar to hybrid teas; climbers, which are sometimes called pillars if less than 8′ tall; creepers—similar to climbers but with weaker stems; tree roses—hybrid teas and floribundas grafted on to slim but sturdy stems about 3′ tall; shrub roses—large, vigorous plants with clusters of old-fashioned, single and semi-double flowers; and miniatures. Roses grow in every part of the country but need special protection in very severe climates. Miniatures also grow indoors.

How to plant. Most roses are sold with bare roots and are planted when still dormant—either early in the spring (from the time the frost leaves the ground until about May 1) or late in the fall (just before the ground freezes). Potted roses are more expensive but can be planted in the spring when they are leafed out and blooming. In either case, get the plants into the ground as soon as possible after you receive them. Remove the wrappings from bare-root, dormant plants and soak roots in a pail of water while you're preparing the soil. Potted plants go directly into the ground.

Plant roses in a spot where they will get at least six hours of full sun a day. Space hybrid teas, hybrid perpetuals, floribundas and grandifloras 18″–24″ apart: climbers, 8′ apart; miniatures, 12″ apart; others, 3′ apart or more.

Planting holes should be 18″ deep and 18″ across. Unless the soil is extremely well drained, place 3″ of coarse gravel in the bottom. Then mix the soil with 1–2 shovelfuls of

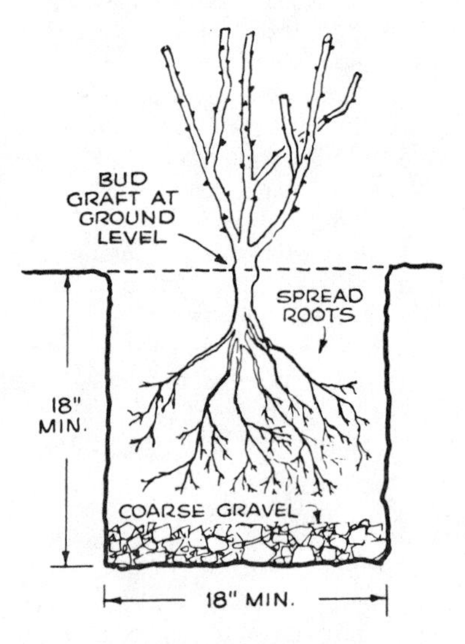

humus (if the soil is very sandy or heavy clay, use twice as much humus) and a handful of balanced commercial rose fertilizer, and place a 6″ layer of this in the hole.

Spread the roots slightly and hold the plant so that the thick bud graft (at the juncture of the canes and root) will be just at ground level when the hole is filled. Then pack soil firmly around the roots until the hole is two-thirds full. Then fill the hole the rest of the way with water.

When the water has soaked in, fill the hole with soil and firm lightly. Then mound soil up around the canes to a height of 8″. If the roses are planted in the spring, don't remove this mound for 10–14 days —until new shoots are growing well. In the fall, leave the mound in place until frost is out of the ground in the spring.

Feeding. Use a balanced commercial rose food such as 8–8–8, scratch it into the soil and water well. The feedings should be made as early in the spring as possible, about May 15 and again about July 15. At each feeding give each plant a handful of fertilizer.

Watering. Roses need the equivalent of a bucketful of water every week through the growing season. Let the water soak into the soil. Don't sprinkle the foliage.

Mulching. It is advisable but not essential to apply a mulch of peatmoss, straw or similar material around roses. It helps to hold in moisture in the summer and to keep down weeds.

Insects and diseases. These can be stopped either by dusting or spraying. The latter is cheaper, neater and less harmful to your nasal passages. Use a combination of 75 per cent phaltan and Isotox according to the manufacturer's directions.

Spraying should start as soon as the roses begin to leaf out in the spring and continue until growth stops in the fall. Spray at seven- to ten-day intervals except during very wet or humid weather, when the frequency of application must be increased.

Pruning. When pruning roses or cutting flowers, make the cuts with sharp shears ¼″ above a leaf bud.

Before growth starts in early spring, prune canes back to live wood; remove weak canes; trim off those that are broken.

During the summer no pruning is necessary except to remove old flowers, broken wood and suckers that shoot up from the rootstock.

In the fall, cut back the tops of very tall plants to a height of about 30″. This helps to prevent them from being whipped about and damaged by wind.

Climbers need very little pruning. Just cut out dead wood and weak and excess canes *after* the plants have bloomed for the first time in the spring. Tie canes to the trellis on which plants are being trained.

Winter care. In the north to protect hybrid teas, hybrid perpetuals, floribundas, grandifloras and miniatures, mound the soil up around the canes to a height of 10″–12″. This protection is not needed in warm climates.

Climbers need no protection except in the extreme north. There they should be wrapped in burlap, straw, etc.

In Ohio and areas with similar climate, tree roses must be wrapped in numerous layers of burlap, straw, canvas, building paper, etc. Farther north they must be bent down and covered with soil.

Shrub roses are generally hardy enough to survive any weather.

Growing miniatures indoors. Miniature roses are not ideal house plants but do pretty well in a sunny south window in a well-ventilated, coolish room. Pot the plants in deep 8″ pots in general-purpose soil to which a little dried manure or bone meal has been added. Keep the soil evenly moist and spray the foliage with water daily. Fertilize every month. Spray as necessary with insecticide-fungicide. Move outdoors during warm weather.

ROSE ACACIA *Robinia hispida.* Bristly, deciduous shrub to 4′ with rose-colored spring flowers. Grows in mild and warm climates.

Grow like locust (which see). The plant suckers freely, especially in light, sandy soil.

ROSEMARY *Rosemarinus officinalis,* old man. Evergreen shrub to 6′ with aromatic leaves and blue flowers. Grows outdoors the year round in warm climates; elsewhere must be brought indoors in winter. Also a house plant.

Outdoors. Propagate by green stem cuttings in early fall. Plant in spring in full sun in average, well-drained soil. Water only in dry weather and then not much. Fertilize every other spring. In the north, the shrub is best grown in a tub which can be moved into a cool, light place indoors in winter, during which period it should be watered sparingly.

Indoors. Start with cuttings in September. Pot in general-purpose potting soil and place in a south window in a cool (about 60°) room. Water when soil feels dry. Fertilize lightly every two months.

ROSE OF SHARON *Hibiscus syriacus.* Sometimes called *althea.* Deciduous shrub to 15′ with large, open flowers in various colors in late summer. Grows in all but coldest climates.

Buy new hybrid varieties. Plant in spring or fall in full sun in average soil that is well drained. Water in dry weather. Fertilize every 1–2 years in spring. Prune out dead wood.

ROYAL POINCIANA *Delonix regia,* peacock flower, flamboyant. Striking, deciduous tree to 30′ covered with red and yellow flowers in spring and summer. Subtropical. See also poinciana.

Plant in spring or fall in sun in average, well-drained soil. Water in dry weather. Fertilize every 2–3 years in early spring.

RUBBER PLANT *Ficus elastica,* India rubber plant. Tree to about 15′ with large, glossy leaves which may be white marked. Best known as a house plant, but also grows outdoors in subtropics.

Indoors. Pot in general-purpose potting soil. Grow in a north window. Keep soil moist and spray or wash off leaves with water every couple of weeks. If plant gets too large, air-layer it. In summer, move outdoors into shade.

Outdoors. Grow in shade in average, moist soil. Demands little attention.

RUBBER VINE *Cryptostegia grandiflora.* Heavy vine to 10′ with thick, evergreen leaves and violet, funnel-shaped flowers in summer. Subtropical.

Propagate by spring stem cuttings. Plant in average soil with humus and general-purpose fertilizer added. Keep moist. Provide sturdy support on which stems can twine. Thin out and cut back lightly in winter. Feed every spring.

RUDBECKIA Gloriosa daisy, coneflower, black-eyed Susan, brown-eyed Susan, thimble flower, golden glow. Plants to 12′ with yellow daisy-like flowers. Mostly perennials but often grown as annuals. Grows in all but hottest climates.

Sow seeds indoors about eight weeks before last frost and shift into a sunny location when weather settles. Grow in average, well-drained soil. Water in very dry weather. Fertilize in spring. Stake plants in windy locations. Plants treated as perennials can be divided in fall or spring.

RUE *Ruta graveolens,* herb of grace. Evergreen perennial to 2′ with strongly aromatic leaves. Grows in temperate climates.

Divide plants and set out in spring or fall in full sun in average soil. Water in dry weather. Cut back rather hard in March or April to encourage new growth. Fertilize at that time.

RUELLIA MAKOYANA Trailing velvet plant. Plant to about 18″ with satiny, silver-veined foliage and red flowers. Grows indoors.

Propagate by young stem cuttings in late spring. Pot in two parts loam, two parts humus and one part sand. Grow in a warm east or west window. Keep soil moist. Stand pots on wet pebbles. Fertilize every two months in summer, fall and winter.

RUTABAGA *Brassica napobrassica.* Large, turnip-like vegetable to 20″, with yellow flesh. Often mistakenly called a turnip. Grows best in cool climates.

Start seeds 110 days before first fall frost may be expected. Grow in full sun in average soil to which general-purpose fertilizer rich in potash has been added. Sow seeds in rows 2′ apart, and thin plants to 1′. Keep weeded. Water in dry spells. Fertilize once or twice during summer. Dig after frost, but don't allow the bulbs to freeze.

SAFFLOWER *Carthamus tinctorius,* saffron. Annual to 2′ with prickly leaves and yellow, thistle-like flowers. Grows almost everywhere.

Sow seeds outdoors when weather is reliably warm. Grow in sun in average soil. Water in dry weather. Fertilize when plants are thinned to 1′.

SAFFRON *Crocus sativus.* Low-growing, cormous plant with lilac-colored, autumn flowers with orange-red stigmas that are dried and used for flavoring. Grows best in cool climates. See also safflower and *colchicum.*

Grow like *crocus* (which see).

SAGE *Salvia officinalis.* Shrubby perennial herb to 2′ with grey leaves and purplish-blue flower spikes. Leaves are used for flavoring. Grows in cool or warm climates. See also salvia.

Divide in early spring or fall and plant in sun in light, average, well-drained soil. Water in dry weather, but sparingly. Fertilize in spring. Pick off leaves or clip the shoots as you want them for flavoring but don't strip the plants.

ST. JOHN'S-WORT *Hypericum.* Different species called rose of sharon or Aaron's beard; goldflower, tutsan; sweet amber; bush broom. Perennials and shrubs, usually evergreen, to 6′, with flowers mainly yellow. Grows in mild and warm climates, although some species can stand a fair amount of cold.

Propagate perennial types by

division; shrubs by green stem cuttings in summer; creeping types by suckers. Plant in spring or fall in light shade in average, well-drained soil. Water in dry weather. Fertilize lightly every year; however, don't expect the plants to live forever.

SALPIGLOSSIS Painted tongue. Annual to 30″ with handsome, petunia-like flowers in mixed colors. Grows almost everywhere.

Sow seeds indoors 8–10 weeks before last frost. Move into a sunny or lightly shaded location in the garden when weather is warm. Grow in average, well-drained soil. Water regularly. Fertilize when young plants are established and once more about six weeks later. Pinch young plants to promote bushy growth. Staking may be necessary in windswept places.

SALVIA Sage. Plants to 8′ with brilliant scarlet, blue or white flowers in summer. Annuals grow almost everywhere. The best of the blue salvias (*S. farinacea* and *S. patens*), both perennials, also grow almost everywhere if treated as annuals. Other perennials are hardy only in warm climates. See also sage.

Sow seeds indoors about eight weeks before last frost. Move seedlings into a sunny location in average soil. Water in dry weather. Fertilize young plants when established. Stake tall blue species. Divide roots of hardy perennials in the spring, when clumps become large.

SANCHEZIA NOBILIS Shrub to 5′ with dark leaves veined with cream, and yellow flowers with red bracts in spring and summer. A house plant; also grows outdoors in warmest climates.

Propagate by greenwood stem cuttings. Plant in average soil in partial shade (or an east window). Indoors, provide as much humidity as possible. Keep moist. Fertilize several times in winter and spring. If grown in pots, repot only when growth is at an obvious standstill.

SAND MYRTLE *Leiophyllum.* Evergreen shrub to 18″ with handsome foliage and small white flowers in spring. Grows in mild and warm climates.

Propagate by layering. Plant in spring or fall in sun or light shade. The soil should contain considerable sand and peatmoss, hold moisture but be well drained. Water in dry spells. Fertilize every 2–3 years.

SANDWORT *Arenaria.* One species sometimes used as a groundcover called Irish moss. Durable, creeping perennials only a few inches tall with pinhead-size flowers. Grows in cool climates.

Divide in spring and plant in sandy soil in sun or shade. Water in dry, hot weather, especially if plants are growing in the sun. Fertilize only if plants need a boost.

SANTOLINA Lavender cotton, ground cypress. Shrubby perennial to 20″ with aromatic, usually silvery-grey foliage. Grows outdoors in warm climates.

Propagate by stem cuttings in spring. Plant in average, well-drained soil in full sun. Water in dry weather. Fertilize very occasionally.

SANVITALIA PROCUMBENS Creeping zinnia. Spreading annual to 6″

with yellow and black ray flowers. Grows almost everywhere.

Sow seeds outdoors where plants are to grow in early spring. Grow in average soil in full sun. Water only in long dry spells. Fertilize lightly when plants are thinned to 15″.

SARCOCOCCA Evergreen shrub to 5′ with lustrous, dense foliage and black berries. Grows in mild and warm climates.

Grow like box (which see). This plant likes a slightly acid soil and regular watering in dry weather. Grow in partial shade.

SASA Genus of bamboos. Grows in warmest climates.

For how to grow, see bamboo.

SASSAFRAS Deciduous tree to 100′ with leaves of several shapes which turn brilliant red in fall. Grows in the east.

Plant young specimens only in spring or fall in sun or light shade. The soil should be of average quality, well drained. Fertilize every 3–5 years, but the trees grow fast anyway. Prune out dead and broken branches as they occur.

SAXIFRAGE Rockfoil, strawberry geranium, Aaron's beard, beefsteak saxifrage, mother of thousands, London pride. Large genus of spreading perennials to 2′, with small flowers in various colors in clusters. Grows in cool climates. Strawberry geranium is a popular house plant.

Outdoors. Divide plants in spring or propagate by pinning the runners down to the soil until they root. Plant in average, well-drained soil in partial shade. Water and cultivate regularly. Fertilize in spring.

Indoors. Pot in general-purpose potting soil, which is kept fairly moist. Place in an east window at 60° or less. Fertilize every 2–3 months.

SCABIOSA Pincushion flower, mourning bride, sweet scabious. Annual or perennial plants to 30″ with pincushion-like flowers in purple, blue, pink and white. Grows almost everywhere.

Annual. Sow seeds indoors about eight weeks before last frost or sow outdoors in full sun or light shade when weather is warm. Grow in average soil which should be limed if acid. Cultivate regularly. Water well during dry spells. Fertilize when young plants are set out or thinned. Keep dead flowers picked off.

Perennial. Propagate by spring divisions or sow seeds in the spring. Divide established plants every three years. Otherwise, treat as above.

SCARBOROUGH LILY *Vallota speciosa.* Bulbous plant to 3′ with clusters of orange-red, funnel-shaped flowers in summer and fall. Grows in warm climates.

This plant does best in well-drained pots in a mixture of two parts loam, two parts humus and one part sand. Plant bulbs right after they have flowered, or about June. The tips of the bulbs should be just above the soil surface. Grow in light shade. Keep moist except in winter, when watering should be reduced but not stopped. Fertilize every six weeks while making growth. Repot only when plant threatens to climb right out of its container.

SCARLET KADSURA *Kadsura japonica.* Dense, dark, evergreen vine to 10′ with red berries in fall. Grows in the southeast.

Propagate by stem cuttings in spring. Plant in sun in average, well-drained soil. Keep watered. Fertilize once or twice in spring and summer. Train to a sturdy support. Prune in early spring.

SCARLET RUNNER BEAN *Phaseolus coccineus.* Vine to 20′ with fiery red flowers and purple pods of edible beans. A perennial but best grown as an annual. Grows almost everywhere.

Grow like bush or pole beans (see bean). Provide a post, wire mesh or trellis for vine to climb on.

SCHIZANTHUS Butterfly flower, fringe flower, poor man's orchid. Bushy annual to 5′ with decorative flowers in many colors in summer Grows best in cool climates.

Sow seeds indoors 8–10 weeks before last frost. Shift into sun or partial shade in average, well-drained soil. Grow in a location protected from wind. Fertilize when plants are established and again about six weeks later. Water regularly and spray foliage with water when the weather is hot. Pinch young plants to make them bush out.

SCILLA Squill, star hyacinth, Spanish bluebell, Spanish jacinth, Cuban lily, sea onion, wood hyacinth. Bulbous plants to 1 with clusters of bell-shaped blue, white or purple flowers in spring. Grows best in temperate climates. Also a house plant.

Plant bulbs in fall 3″–4″ deep and about that far apart. Grow in average, well-drained soil with a little humus added. Fertilize every other year in late winter. Scillas grow in sun but are outstanding in shade, even under evergreen trees. Divide clumps about every four years.

To grow scillas indoors, see How to Force Bulbs for Indoor Bloom.

SEA BUCKTHORN *Hippophae rhamnoides.* Thorny, deciduous shrub to 25′ with willow-like leaves and large clusters of orange berries in fall. Grows almost everywhere.

Propagate by suckers or by layering. Plant in spring or fall in sun in any kind of soil, even sand. No attention is called for Male and female specimens must be planted rather close together however, to ensure fruit.

SEA GRAPE *Coccoloba uvifera.* Evergreen tree to 15′ with leathery leaves and edible fruits like a bunch of purple grapes. Grows in warmest climates.

Plant in spring or fall in sun. Soil should be well drained, sandy Water in dry weather Fertilize every 2–3 years.

SEA HOLLY *Eryngium*, eryngo. Perennials to 7′ with spiny leaves and strange, metallic-blue flowers in summer. Grows in temperate climates.

Divide plants in spring, or sow seeds immediately after they are harvested for flowers the next year Grow in sun in average soil. Water regularly Fertilize in spring. Spray with captan or fermate to prevent rot and leafspot.

SEDUM Different species known as stonecrop, wall pepper, Christmas cheer, Boston bean, burro's tail, donkey tail. Succulents of many different sizes and shapes. Many species grow almost everywhere—even in very cold climates; others grow only in warmest climates. Many sedums also grow indoors.

For how to grow see cactus, desert. Keep indoor plants at 60°–65°.

SELAGINELLA KRAUSSIANA Club moss. Creeping perennial like a moss. Grows as a groundcover in subtropics. Also a house plant.

Propagate any time by stem cuttings inserted in average soil where plants are to grow. Give shade and regular watering. Indoors, spray with water often.

SEMPERVIVUM Houseleek. Stemless, rosette-shaped succulents to 1′ tall and 6″ diameter, which spread widely and produce white, yellow, pink or purple summer flowers. Except for a few tender species, *sempervivum* grows almost everywhere. Also a house plant.

For how to grow, see cactus, desert.

SENSITIVE FERN *Onoclea sensibilis.* Four-foot fern with coarse, triangular fronds. Very sensitive to frost. Grows in temperate climates east of the Rockies.

For how to grow, see fern. This plant grows in partial shade or sun but requires a rather heavy, rich, humusy soil that holds moisture well.

SEQUOIA Redwood, big tree. Giant conifers to 350′ in the wild but much smaller in cultivation. Grows on the West Coast.

Plant in spring or fall in sun or partial shade in average soil that is well drained. Water deeply about once a month in dry weather.

SERISSA FOETIDA Shrub to 3′ with box-like leaves and profuse white flowers in summer. A house plant; also grows outdoors in warmest climates.

Propagate by semi-hardwood stem cuttings. Plant in average, well-drained soil in partial sun (or an east or west window). Keep soil moist. Fertilize in winter and spring; more often if in a pot. Prune after flowering.

SHADBUSH *Amelanchier,* shadblow, serviceberry, Juneberry. Deciduous trees or shrubs to 45′ with white flowers in early spring and purple fruits sometimes used for jelly. Grows in temperate climates.

Plant in early spring or fall in a sunny location in average, well-drained soil which is limed occasionally if acid. Water in long, dry spells. Fertilize every 3–4 years in the spring.

SHALLOT *Allium ascalonicum.* One-foot member of the onion family with bulbs separated into cloves like the garlic. Grows almost everywhere except in hot, dry regions.

Grow like onion (which see). Plant the cloves like onion sets. Young plants may be pulled like scallions or mature bulbs harvested in fall.

SHASTA DAISY *Chrysanthemum maximum.* Two-foot perennial with double or single, white, daisy flowers. Grows best in cool climates.

Divide in spring and plant in deep, well-drained, humusy soil in sun or partial shade. Fertilize in early spring and twice more at monthly intervals. Water regularly. Pinch stem ends to promote bushy growth. Spray with a combination insecticide-fungicide.

SHEPHERDIA Buffaloberry, wild oleaster. Deciduous shrubs to 18′ with silvery leaves and yellow or red, sour berries used in jelly. Grows in mild and coldest climates.

Plant male and female specimens if you want fruit. Plant in spring in sun in average, well-drained soil. These are very tough plants and need water only in extreme drought. Fertilize every 1–2 years in spring.

SHRIMP PLANT *Beloperone guttata.* Evergreen shrub to 3′ with pendulous, salmon-pink bracts that look like shrimps. A house plant. Also grows in subtropics.

Indoors. Propagate by stem cuttings in spring. Pot in general-purpose potting soil with a little extra humus. Grow in a south window. Water when soil feels dry. Fertilize every three weeks. To promote bushiness, pinch stem ends. Discard leggy plants after taking cuttings from them.

Outdoors. Plant in average, well-drained soil in partial shade. Handle as above but fertilize only a couple of times a year. Cut back stems when flowers die.

SIDALCEA False mallow, miniature hollyhock. Perennials to 4′, very much like small hollyhocks. Grows mainly west of the Rockies.

Divide plants every three years in spring or fall and plant in sun in average soil improved with humus. Water and cultivate regularly. Fertilize in spring. Cut back flowering stems to encourage a second showing.

SILENE Campion, catchfly, cushion pink, Indian pink. Large family of annuals and perennials to 3′ with sticky stems and small white, pink and red flowers. Grows almost everywhere.

Sow seeds of annuals in fall where the plants are to grow. Divide perennials in spring. Grow in sun in average, sandy soil. Fertilize in spring. Water in dry weather.

SILK TASSEL *Garrya.* Evergreen shrubs to 8′ with green and grey leaves, yellow or green flower tassels and clusters of purple fruit. Grows from northern California northward.

Plant male and female specimens for fruit. Plant in spring or fall in sun or light shade. Soil of average quality must have good drainage. Water deeply every month in dry weather. Fertilize in mid-fall. Prune after fruiting.

SILK TREE *Albizzia julibrissin.* Also incorrectly called mimosa. Flat-topped, spreading, deciduous tree to 40′ with lacy foliage and ball-like clusters of pink flowers in summer. Grows in warm climates.

Plant small specimens only in spring or fall in average soil in the sun. Watering and fertilizing are not required although an occasional dose of the latter helps. Prune new growth any time after flowering. In the south, the tree may be attacked by a disease for which there is no cure.

SILK VINE *Periploca.* Deciduous vine to 40′ with waxy foliage, greenish-brown, summer flowers and seed pods with silk filaments. Grows in mild and warm climates.

Propagate by layering. Grow in average soil in the sun. Water in very dry weather. Fertilize lightly in spring. Provide a strong support and train stems to it in order to keep them from tangling. Prune back hard in early spring.

SILVER BELL TREE *Halesia,* snowdrop tree, tisswood. Deciduous tree to 90′ with a profusion of hanging, white, spring flowers. Grows best in warm climates.

Plant in early spring or fall in a partially shaded place protected from wind. Don't plant in frost pockets. Provide rich, well-drained soil. Water in long dry spells. Fertilize every 3–4 years.

SILVER LACE VINE *Polygonum auberti,* fleece flower. Dense, deciduous vine to 25′ with masses of white flowers in summer. Grows in all but coldest climates.

Propagate by stem cuttings. Plant in average soil in full sun. Water in long dry spells. Fertilize lightly in the spring. Tie stems to a sturdy trellis. Pinch stem ends to promote bushing out. Cut back hard and thin out the stems in the fall. Very cold weather may kill the vine to the ground, but it should make a strong comeback the next year.

SILVER TREE *Leucadendron argenteum.* Evergreen tree to 40′ with dense, silky, silvery foliage. Grows in warmest climates.

Plant young specimens in spring or fall in a sunny place protected from the wind. The soil should be of average quality, very well drained. Water deeply about once a month in dry weather. Fertilize every 3–4 years in the spring.

SKIMMIA Ornamental evergreen shrubs to 5′ with white flowers and red berries. Grows in mild and warm climates.

Propagate by stem cuttings. Plant in spring or fall in shade in a good, well-drained loam containing some sand. In the case of some species, both male and female specimens are needed to ensure fruit. Water in dry weather. Fertilize in spring. Prune in winter.

SMILAX *Asparagus asparagoides.* Perennial vine to 20′ with oval-shaped leaves much used by florists. Grows indoors. Other plants bearing the botanical name of *smilax* are greenbriers of little use to gardeners.

Grow like asparagus fern (which see).

SMITHIANTHA Temple bells. Formerly identified as *naegelia.* Plants growing from rhizomes to 2′ with cream to red, bell-shaped flowers and pretty leaves marked with red or purple or covered with red or purple hairs. A house plant.

Propagate by division of the rhizomes. Pot rhizomes in summer on their sides and ½″ below the soil surface. The soil should be a sterilized mixture of equal parts of loam, humus and sand. Set pots in a light, warm place and water only a little until growth starts. Then move to an east or west window and set pots on wet pebbles. Keep soil evenly moist. Fertilize every two weeks. When flowering stops, stop feeding and gradually reduce water until foliage dies. Then store rhizomes in their pots in a dark, cool place until summer.

In areas where the temperature stays at 70° or more during the summer, smithianthas can be grown outdoors in pots in the above manner.

SMOKE TREE *Cotinus*, chittamwood. Often identified as *Rhus cotinus*, Deciduous shrubs or trees to 20′ with masses of airy flower plumes in summer and beautiful fall foliage. Grows in temperate climates.

Plant in early spring or fall in a sunny spot. The soil should be of average quality and well drained. Water newly planted specimens well; but once plants are established, little is needed. In the north, give some shelter from wind. Fertilize every 3–5 years.

SNAKE CACTUS *Nyctocereus serpentinus*. Columnar desert cactus to 8′ with red and green spines and white flowers at night. Grows in hot, dry climates; also indoors.

For how to grow, see cactus, desert.

SNAKE PALM *Hydrosme rivieri*, Devil's tongue, black sacred lily of India. Also identified as *Amorphophallus rivieri*. Tuberous plant to 4′ with huge leaves resembling a snake's skin and a flower spathe like a calla lily. Grows in subtropics.

Plant tubers in early spring in equal parts loam, humus and sand. Set the tubers 3″ deep and space at least 3′ apart. Single plants can be easily grown in large pots or tubs. Grow in light shade. Water when soil feels dry. Fertilize every two months while plant is growing. Stop feeding and watering when foliage begins to yellow. If you live where the ground freezes or is wet in winter, tubers should be lifted and stored in a cool, dark place over winter.

SNAKE PLANT *Sansevieria*, bowstring hemp. Upright foliage plant to 5′ with thick leaves that are striped or banded yellow or light green. Mainly a house plant, although it also grows outdoors in subtropics.

Propagate by 4″ lengths of leaves inserted, right end up, 2″ deep in moist sand. Snake plants with variegated leaves, however, must be divided. Pot in general-purpose potting soil in containers with a thick layer of coarse drainage material. Grow in a north window. Water thoroughly when soil feels dry. Fertilize every three months. Move outdoors into shade in summer.

SNAPDRAGON *Antirrhinum*. Erect plants to 3′ with spikes of flowers in many colors. Perennials but usually treated as annuals. Grows almost everywhere. Sometimes grown indoors.

Outdoors. Sow seeds indoors 8–10 weeks before last frost. Shift into full sun or partial shade in average, well-drained soil to which a little humus has been added. Fertilize when young plants are established and again before flowering starts. Water in dry weather. Pinch young growth once or twice to promote bushiness. In warm climates, when growth dies down in the fall, cover plants with several inches of salt hay. They should carry over the winter.

Indoors. Start seeds outdoors in August and plant in individual pots in general-purpose potting soil. Move indoors before frost and keep in a very cool, south window. Water when soil feels dry.

SNOWBERRY *Chiogenes hispidula.* Creeping, evergreen shrub only a few inches high with white flowers and white berries. Grows in cold and cool climates.

Divide in spring and plant in shade in moist, humusy, acid soil. Water in dry weather. Fertilize if plants need a boost.

SNOWDROP *Galanthus.* Bulbous plants to 1′ with tiny, white, bell-like flowers in late winter. Grows best in cool climates. Also indoors.

Plant in partial shade in moist, humusy soil. Set bulbs 3″ deep and 3″ apart. Don't let the bulbs stay out of the ground any longer than possible. Once established, leave them alone. To grow snowdrops indoors, see How to Force Bulbs for Indoor Bloom.

SNOWFLAKE *Leucojum,* summer snowdrop. Bulbous plants to 1′ with nodding flowers, mostly white, in spring or fall. Grows almost everywhere.

Plant the bulbs in the fall in average, well-drained soil in light shade. The spring-flowering species go 3″ deep and that much apart; the autumn-flowering, 4″ deep and 3″ apart. Water in very dry weather.

SNOW ON THE MOUNTAIN *Euphorbia marginata.* Annual to 15″ with white-margined leaves and white flowers. Grows almost everywhere.

Sow seeds outdoors when soil can be worked. Grow in sun or light shade in almost any soil. Water in long dry spells. Fertilize when plants are well established.

SOLANDRA One species known as cup of gold or chalice vine. Strong evergreen vines to 30′ with huge yellow or white flowers. Subtropical.

Plant in average to lean soil in spring or fall. Grow in sun. Protect from wind. Water well in fall and winter; reduce supply in late spring and summer. Fertilize in October or November. Tie stems to a strong support. Prune in late spring, after flowering. Plants that get too large do not bloom well.

SOLOMON'S SEAL *Polygonatum.* Perennials to 8′ with arching stems from which small, white flowers hang in the spring. Grows in temperate climates.

Divide in spring or fall and plant in average soil which is kept moist and fertilized in spring. Grow in partial shade. Cultivate shallowly.

SORBARIA False spirea. Deciduous shrubs to 18′ with clusters of white summer flowers. Most species grow in mild climates but one is hardy in the north.

Propagate by suckers. Plant in spring or fall in a place where the plants will not crowd others by their fast-spreading habits. Grow in partial shade in average soil that holds moisture. Water in dry weather. Cut old growth to the ground in the spring and apply a little fertilizer at that time.

SORREL *Rumex acetosa,* sourgrass. Perennial to 3′ with long, thick, upright leaves used in salads and as spring greens. Grows in temperate climates.

Sow seeds outdoors in the sun any time until July. Transplant in early fall to a sunny spot in average soil. Fertilize then lightly and yearly thereafter in early spring. Water in dry weather.

SOURWOOD *Oxydendrum arboreum,* sorrel tree. Deciduous tree to 40′ with clusters of lily-of-the-valley-like white flowers in midsummer and deep-red fall foliage. Grows in mild and warm climates.

Plant in early spring or fall in a sunny situation. Grow in average, well-drained soil which is slightly acid. Water in long dry spells. Fertilize every 2–3 years. Keep dead wood cut out.

SOUTH AMERICAN JELLY PALM *Butia capitata.* Palm tree to 24′ with bluish fronds arching outward and acid fruits sometimes used for jelly. Can withstand temperatures to about 15°.

For how to grow, see palm.

SPANISH BROOM *Spartium junceum,* weaver's broom. Deciduous shrub to 8′ with rush-like stems and yellow flowers in clusters. Grows in warmest climates. See also broom and *genista.*

Grow like broom (which see).

SPANISH SHAWL *Schizocentron elegans.* Creeping plant only a few inches tall with reddish-purple flowers and small leaves. Grows indoors.

Propagate by stem cuttings in summer. Pot in general-purpose potting soil. Grow in a warm east or west window. Keep soil moist. Spray foliage with water frequently. Fertilize every three months. After flowers die, cut stems back 50 per cent.

SPARAXIS Wand flower, harlequin flower. Plants growing from corms to 18″ with short spikes of yellow, purple, red or multi-colored flowers in late winter. Grows in warm climates; also a house plant.

Grow like *ixia* (which see) outdoors. Indoors, grow like *freesia* (which see).

SPATHIPHYLLUM Flame plant. Perennial to 1′ with long, narrow leaves and long-lasting, white or yellowish flowers. A house plant; also grows outdoors in subtropics.

Divide roots in February and plant in two parts loam, two parts humus and one part sand. Grow in a warm east or west window (or in partial shade outdoors). Stand pots on moist pebbles. Keep soil moist and fertilize every two months except November through January, when feeding is stopped and water supply reduced.

SPIDER FLOWER *Cleome.* Annual to 5′ with large clusters of white, pink or rosy-purple flowers. Grows almost everywhere.

Sow seeds where plants are to grow in full sun in early spring. Grow in average, well-drained soil. Water in dry spells. Fertilize when plants are thinned to 18″. Remove seed pods before they open, otherwise plants will self-sow freely.

SPIDER LILY *Hymenocallis calathina,* Peruvian daffodil, basket flower. Often identified as *ismene.* Bulbous plant to 2′ with large, airy, tubular, white flowers in summer. Grows almost everywhere.

Plant bulbs in the spring after danger of frost is past. Set them 3″ deep and 6″ apart in average, well-drained, sandy soil in a sunny location. Water copiously in dry weather. Fertilize after growth is well up. In the fall, dig up and store the bulbs in a dry, cool place. Don't remove the roots. Save little offset bulbs for planting out in a nursery bed in the spring.

SPIDER PLANT *Chlorophytum elatum.* Incorrectly identified as *anthericum.* Perennial to 3′ with long leaves marked white or yellow and lax stems with little plants at the ends. A house plant.

Propagate by layering the stems. Pot in general-purpose potting soil. Grow in a north, east or west window. Keep soil moist. Fertilize every three months. Prune as necessary to keep under control.

SPIDERWORT *Tradescantia,* snake grass. Perennials to 3′ with attractive foliage and long-lasting flowers in purple, blue, white or pink. Grows mainly in mild and warm climates. See also wandering Jew.

Propagate by division in the spring or by stem cuttings. Grow in light shade or sun in average soil which should be kept fairly moist. Fertilize in the spring.

SPINACH *Spinacia.* Leafy vegetable to about 1′. It grows almost everywhere but most varieties go to seed in hot weather. One variety that likes heat is New Zealand spinach.

Spinach is happy in full sun in average soil. At planting time, mix in a little general-purpose fertilizer rich in nitrogen. Sow seeds in rows 1′ apart as early in the spring as the soil can be worked. Make succession sowings at fortnightly intervals until about 50 days before hot weather sets in. Thin plants to about 3″. Keep weeded. Water in dry spells. Cut the entire plant, not just a few leaves, when it is ready to eat. Additional sowing of seeds can be made in August for a fall crop.

New Zealand spinach seeds should be soaked in water for two days before they are sown in rows 2′ apart. Space seeds about 1½″ apart and thin to 8″. Small plants are easily transplanted. Water, fertilize and weed as above. The young stem tips can be picked repeatedly throughout the summer.

SPIREA *Spiraea.* Different species called bridal wreath, meadowsweet, queen of the meadow, bridewort, hardhack. Deciduous shrubs to 6′ with a profusion of white or pink flowers—some species in spring, others in summer. Grows almost everywhere, although most species cannot tolerate severe cold.

Propagate by green stem cuttings or by layering the stem ends of arching types of spirea. Plant in spring or fall in full sun. Grow in average soil which retains moisture but is well drained. Water in dry weather. Fertilize in early spring. Prune early-flowering types after they bloom by cutting old stems to the ground. Cut back stems of summer-flowering types one-third to one-half in late winter.

SPRUCE *Picea.* Handsome conifers to 180′ with four-sided needles and hanging cones. Grows best in cool, moist climates.

Plant in spring or early fall in sun or partial shade. Avoid exposure to strong winds. Grow in average, well-drained but somewhat moisture-retentive soil. Water in long dry spells. Fertilize newly planted trees for a couple of years. Spray with malathion in early spring and fall if woody growths, known as galls, appear on shoots. If necessary to prune the trees, do so in July or August.

SQUASH *Cucurbita.* There are many shapes and varieties for summer, fall and winter use. Grows almost everywhere.

Summer squash. Squashes of all types grow in average soil which should be improved at planting time with humus and fertilizer. Full sun is required. Sow summer squash seeds 1″ deep in rows 4′ apart as soon as all danger of frost is past. Thin plants to stand 3′ apart. Keep weeded and water heavily in dry spells. Dust every week until flowering starts with rotenone. At the end of the season, burn all debris.

Winter squash. Handle like summer squash except plant a half-dozen seeds in a 1′ circle (hill) and thin to about two plants. The hills should be 6′ apart in all directions. Harvest the squash before frost, leaving part of the stem on the fruit. Keep in a dry place at about 70° for two weeks; then store at about 45°.

STAGHORN FERN *Platycerium.* Magnificent, tree-dwelling fern with fronds resembling antlers and sometimes extending as much as 8′. Grows indoors; also outdoors in warmest climates.

All species can be propagated by suckers except *P. grande.* Buy this one. The fern is "planted" by wrapping the roots in sphagnum moss or osmunda fiber, into which a little charcoal and bone meal have been mixed, and wiring them to a stick of wood. Grow in a north window or outdoors in shade. Keep roots moist and spray the foliage with water every day or two except in winter, when the plants are resting and should be sprayed no more than once a week. During winter, give a nighttime temperature of about 60°; a daytime temperature of about 68°. At other times of the year temperatures should be somewhat higher. While plants are making growth, feed every 4–6 weeks with a little dilute liquid fertilizer.

STAPELIA Carrion flower, starfish flower. One species called toad cactus. Succulents to about 9″ with extraordinarily mixed-colored, five-pointed flowers which smell like rotten meat. Grows in hot, dry climates; also indoors.

For how to grow, see cactus, desert. Water moderately in summer and provide light shade at that time. In winter, water sparingly. Feed lightly at the start of the growing season in spring.

STAR CACTUS *Astrophytum.* Different species called bishop's cap, sand dollars, goat's horn. Desert cacti to 15″ with a few prominent ribs and yellow flowers. Grows in hot, dry climates; also indoors.

For how to grow, see cactus, desert. Plant shallowly in limed soil.

STAR JASMINE *Trachelospermum jasminoides,* Confederate jasmine. Evergreen vine to 20′ with clusters of fragrant, white flowers in spring and summer. Grows in warm climates. Also a house plant.

Outdoors. Propagate by layering. Plant in spring or fall in moist, light soil with humus added. Fertilize in spring and fall. Grow in partial shade, although in cooler climates it likes full sun. Tie stems to a sturdy support. Cut out dead and damaged wood and prune to control shape in the fall.

Indoors. Pot in general-purpose potting soil with humus added. Grow in a warm, south window and train stems on cords or a light lattice. Water when soil feels dry. Fertilize every two months from spring through early fall. Move outdoors in summer.

STAR OF BETHLEHEM *Ornithogalum.* One species called chincherinchee. Bulbous plants to 3′ with variously shaped, white, spring flowers. Most popular as house plants, but several hardy varieties grow outdoors in temperate climates.

Indoors. Plant bulbs in the fall 1″ deep and three or more to a large pot filled with general-purpose potting soil. Keep in a light, cool place until flower buds are well up and beginning to swell; then move into a south window. Water regularly from time of planting. Apply liquid plant food when pot is moved to a bright window and again before flowering starts. After flowering, gradually withhold water. Store bulbs in their pot in a cool, dry place.

Outdoors. Plant bulbs 3″ deep and about 6″ apart in the fall. Grow in average, well-drained soil in the sun. The plants need no care except water in dry spells. Dig up the bulbs and separate them frequently.

STAR OF TEXAS *Xanthisma texanum.* Annual to 4′ with canary-yellow ray flowers. Grows in dry climates.

Sow seeds in early spring where the plants are to grow. Grow in sun in average soil. No watering or fertilizing are required, though plants certainly won't resent it if you give them a little attention.

STAUNTONIA HEXAPHYLLA Evergreen vine to 40′ with fragrant, white, bell-shaped flowers in the spring and edible red fruits in fall. Grows in warm climates.

Propagate by layering. Plant in spring or fall in good, well-drained soil with a large amount of humus. Don't let soil dry out at any time. Fertilize in spring, summer and fall with nitrogen-rich general-purpose plant food. Protect from wind and grow in partial shade. Provide stout support for vine to climb on. Prune in early spring.

STEPHANANDRA Deciduous shrubs to 8′ with white flower clusters much like spirea. Grows in mild and warm climates.

Propagate by stem cuttings. Plant in spring or fall in sun in average, well-drained soil. Water in dry weather. Fertilize in spring. Prune in late winter. Plants may get killed to the ground in severe weather, but usually put up new growth from the roots.

STERNBERGIA LUTEA Winter daffodil, yellow autumn crocus. Bulbous plant to 1′ with yellow, crocus-like flowers in fall. Grows best in temperate climates.

Plant the bulbs 4″ deep and about that much apart in average, well-drained soil in early August. Grow in sun in a warm, sheltered location. Fertilize lightly about every two years before growth starts in late summer. If necessary to transplant, do so in the spring after foliage dies down (it lasts through the winter).

STEWARTIA Mountain camellia, silky camellia. Sometimes spelled *Stuartia.* Deciduous shrubs or trees to 15′ with cup-shaped, white flowers in summer and brilliant fall foliage. Grows in warm climates.

Plant in spring or fall in light shade or sun. The soil should be rich in loam and humus, well drained but moisture-retentive. Water in dry weather. Fertilize every 2–3 years in the spring.

STIGMAPHYLLON CILIATUM Golden vine, butterfly vine. Evergreen vine to 12′ with bright green foliage and profuse golden flowers in summer. Subtropical.

Propagate by stem cuttings over heat in late winter. Plant in spring or fall in rich, humusy soil which is kept evenly moist. Fertilize and prune in early spring. Train on a light trellis. Needs partial shade and protection from wind.

STOCK *Matthiola incana.* Biennial usually handled as an annual. The best-known type grows to 2′ and has large, fragrant flower heads in many colors. Night-scented stock has single lilac flowers.

Sow seeds indoors 10–12 weeks before last frost. Keep in a cool, sunny window. Don't overwater. Transplant into individual pots when seedlings have 4–6 leaves. Then move outdoors when weather is reliable. Grow in full sun in deeply dug, well-drained soil. Water regularly. Fertilize young plants when established and monthly thereafter. To produce largest flower spikes, pinch out side shoots. Stock needs cool weather to bloom well.

STOKESIA Stoke's aster, cornflower aster. Much-branched perennial to 18″ with large, cornflower-like, blue flowers. Grows in mild and warm climates.

Divide plants in spring or fall and plant in a sunny spot in light, well-drained soil. Water regularly. Fertilize in early spring.

STONE CRESS *Aethionema.* Perennials to 1′ with clusters of mainly pink flowers. Rock-garden plants. Grows in cold and mild climates.

Sow seeds outdoors in a seedbed in spring when weather is warm, or divide plants in early spring and plant in sun in average, very well-drained soil containing considerable sand and some lime. Water in dry weather. Fertilize lightly in spring. After flowering, prune back somewhat. Cover with salt hay or evergreen boughs in very cold climates.

STRANVAESIA Evergreen shrub to 20′ with white spring flowers, bronze-purple fall foliage and red berries. Grows best in mild, moist parts of the northwest.

Propagate by half-ripe stem cuttings in summer. Plant in spring or fall in average, well-drained soil. Water in dry spells. Fertilize every 2–3 years in spring. Prune after flowering.

STRAWBERRY *Fragaria.* This favorite, low-growing perennial fruit grows almost everywhere if you select the right varieties. Cultivated berries are classified as June-bearing types—those that bear one crop of berries in the spring—and everbearers—those that bear in spring and again in fall.

Buy plants or propagate by the runners. Plant as early as possible in spring. Strawberries need full sun. The soil should be thoroughly spaded and pulverized; well drained. Mix extra humus into average soil; also mix in well-rotted or dried manure and a little bone meal. General-purpose fertilizer may also be used but must be worked and watered into the soil about a week before the plants are set out. If plants cannot be set out when you receive them, water the roots well and wrap in paper, or heel the plants in. Make planting holes wide enough so that the roots, which should be sheared to a length of 4″, can be spread out. Set plants so that crowns are just level with the soil surface. Firm soil well and water thoroughly.

Strawberries are spaced and grown in various ways, but for the average gardener, the easiest system yielding the best fruit is the hill system. Using this, you space the plants 1′ apart in rows 2′ apart and thereafter keep the runners clipped off as they develop.

During the first spring, keep blossoms picked off so that berries will not form. Through spring and summer, cultivate the soil very shallowly—just enough to keep the surface crumbly and free of weeds—every fortnight. Water deeply in dry weather. At midsummer, if you planted everbearing strawberries, allow blossoms to set so that you will have fall fruit. In September, mix general-purpose fertilizer into the soil at the rate of 1 lb. per plant. Later, when the ground freezes, cover the plants with 3″ of straw to protect them against heaving.

In the second spring, when frost danger is past, remove some of the straw from over the plants but not from alongside. Leave that to hold in moisture, discourage weeds and to keep the berries out of the dirt. Keep runners clipped off. Allow blossoms to form and develop into fruit.

During the second year you will get two crops of berries from everbearing types—one in spring and another in fall. After the second, tear up the plants and start a new batch the next spring.

The June-bearing strawberries should also be ripped out during the second year—in this case, after they have finished bearing their spring crop. However, if the plants seem very healthy, you can "renew" them in the following manner: After the plants have stopped bearing, remove the mulch and cut all foliage from the plants (but be careful not to cut the crowns). Burn the foliage. Pull up weak and surplus plants. Cultivate regularly during the summer, and water as necessary. In early fall, apply fertilizer. Then handle as you did the previous fall, winter and spring. After the plants have borne their second crop in the following spring, however, they should be removed.

STRAWFLOWER *Helichrysum bracteatum.* Familiar everlasting to 3′. An annual, it grows almost everywhere.

Sow seeds in full sun in average soil when weather is warm. Water in dry weather. Fertilize when plants are thinned to 15″. Cut flowers before they open all the way; strip off leaves, and hang upside down in a shady place until dry.

STREPTOCARPUS Cape primrose. Plants to about 2′ with leaves near the ground and long stems bearing tubular flowers in many colors. A house plant.

Sow seeds indoors at any time of year or propagate by leaf cuttings. Pot in equal parts of loam, humus and sand. Grow in an east or west window. Stand pots on wet pebbles. Keep soil moist. Fertilize with dilute liquid plant food every 2–3 weeks. Repot annually.

STROBILANTHES Plants to 30″ with handsome foliage and clusters of flowers in purple, blue or white. A house plant.

Propagate by stem cuttings in spring. Pot in two parts loam, two parts humus, and one part sand. Grow in a warm east or west window. Stand pots on wet pebbles and mist foliage daily with water. Keep soil moist. Fertilize every 2–3 months. Start with new plants every year.

STYRAX JAPONICA Japanese snowbell, Japanese snowdrop. Deciduous trees to 30′ with small, fragrant, white flowers in spring or early summer. Grows in all but coldest climates.

Plant in early spring in the sun. The soil should be of average quality and well drained. Water in dry weather. Fertilize every 2–3 years in the spring. Prune young plants in winter to develop tree shape; without pruning they are likely to become shrubby.

SUCCULENT One of many different types of plants which store water and grow, as a rule, in dry locations. Specific popular succulents are listed under their own names.

Grow like desert cactus (see cactus, desert).

SUMAC *Rhus.* Deciduous shrubs and trees to 30′ with brilliant red fall foliage and often handsome fruits. Grows almost everywhere.

Plant in spring or fall in sun in any soil. No other care is called for. These are showy plants in the fall but of very little value in other seasons.

SUMMER HYACINTH *Galtonia candicans,* cape hyacinth. Bulbous plant to 3′ with clusters of pendulous, white flowers in summer. Grows almost everywhere.

Plant in late spring 3″ deep and 1′ apart in rich, well-drained soil in the sun. Water in dry weather. Fertilize when growth starts and again before flowering. In the fall, dig up the bulbs, dry off and store in a cool, dry place in trays. In warm climates the bulbs can be left in the garden. Don't count on bloom every year.

SUMMER SAVORY *Satureia hortensis.* Annual herb to 18″ with pink flowers and leaves used for flavoring. Grows almost everywhere.

Sow seeds indoors 8–10 weeks before last frost. Transplant into average soil in a sunny location. Water regularly. Fertilize when plants are established.

SUNFLOWER *Helianthus.* Annuals and perennials to 12′ with huge yellow flowers. Grows almost everywhere.

Sow seeds of annuals in early spring. They should be ½″ deep in average soil that is deeply dug, limed and fertilized. Propagate perennials by division in early spring. Sunflowers need full sun. Water regularly and fertilize several times during spring and summer.

Cultivate frequently. Small varieties should be pinched back when they are young to promote bushier growth. Tall varieties need to be staked. Burn the plants in the fall.

SUPPLEJACK *Berchemia scandens.* Open, deciduous vine to 20′ producing small fruits which turn from red to black. Grows in warm climates.

Grow from seeds or stem cuttings in spring. Plant in spring or fall in average, well-drained soil in full sun. Water in dry weather. Fertilize in spring. Train twining stems up a light trellis. Prune in spring.

SWAN RIVER DAISY *Brachycome iberidifolia.* Annual to 18″ with lacy foliage and blue, white or rose, daisy-like flowers. Grows almost everywhere.

Sow seeds indoors about eight weeks before last frost and move outdoors into a sunny location in average soil. Water in dry weather. Fertilize when young plants are established.

SWEET ALYSSUM *Lobularia maritima,* snowdrift. Annual to 1′ with masses of white, violet or pink flowers. Grows almost everywhere; also indoors. See also *alyssum.*

Outdoors. Sow seeds indoors eight weeks before last frost, or sow outdoors where plants are to grow after last frost. Sweet alyssum needs average soil, sun, moisture in dry weather. Fertilize when young plants are established or thinned to 6″ apart. The plants also do well in boxes or pots, but must be watered regularly and fertilized more often.

Indoors. Sow seeds outdoors in August. Pot up in general-purpose potting soil. Move indoors before frost. Keep in a cool, sunny window. Water when soil feels dry. Fertilize lightly about once a month.

SWEET BAY *Laurus nobilis,* bay tree, Grecian laurel. Evergreen to 40′ but normally held to smaller size. Grows in warmest climates. Also possible as a house plant if you have a very cold, shady sun porch.

Plant in spring in sun in well-drained soil containing considerable humus and sand. Water regularly in spring, summer and fall but very little in winter. Give a substantial dose of plant food every spring. Clip to shape (sweet bay takes kindly to shearing and is grown in many shapes) after new growth has been made in early summer. Spray with malathion to control scale.

SWEET CICELY *Myrrhis odorata.* Perennial herb to 3′ with fern-like, fragrant leaves used for flavoring. Grows in temperate climates.

Divide in spring or fall, or sow seeds outdoors in those seasons. Grow in sun or partial shade in average soil. Water in dry weather. Fertilize in spring.

SWEET FERN *Comptonia asplenifolia.* Deciduous shrub to 3′ with fern-like foliage which is fragrant when crushed. Grows in the east.

Propagate by layering. Plant in spring or fall in the sun in poor, dry soil.

SWEET FLAG *Acorus calamus.* Marsh plant with thick, grass-like foliage to 2′ and with fleshy fragrant roots. Grows almost everywhere.

Divide roots in spring and plant in rich soil that remains damp at all times. Plant will also grow in shallow water.

SWEET GUM *Liquidambar styraciflua.* Very handsome, symmetrical deciduous tree to 120′ with star-shaped leaves which turn brilliant red in the fall. Grows in all but extreme climates.

Plant in early spring or fall in sun or partial shade. Grow in rich, moisture-retentive, humusy soil. Water in long dry spells. Fertilize every 3–5 years.

SWEETLEAF *Symplocos tinctoria.* Evergreen tree to 25′ with clusters of fragrant, yellowish-white flowers in spring and small, orange-brown fruits. Grows in warm climates.

Plant in spring or fall in sun or partial shade in moist, average soil. Plant young specimens only. Water in dry spells. Fertilize in late winter every 2–3 years.

SWEET MARJORAM *Majorana hortensis.* Sometimes identified as *Origanum majorana.* Perennial herb to 2′ grown (usually as an annual) for its leaves, which are used for flavoring. Grows almost everywhere.

Sow seeds outdoors when frost danger is past. Grow in sun in average, well-drained soil. Water in dry weather. Fertilize when plants are thinned to 9″ apart in the row. Pot up some of the plants in the fall and bring them indoors to grow in a cool, sunny window.

SWEET OLIVE *Osmanthus fragrans,* tea olive. Evergreen shrub or tree to 25′ with attractive, glossy foliage and fragrant white flowers in early spring. Grows outdoors in warmest climates; also indoors.

Outdoors. Propagate by half-ripe stem cuttings in late summer. Plant in fall or spring in a spot not receiving midday sun and out of the wind. Grow in average, slightly acid soil and water regularly. Fertilize once a year. Prune after flowering.

Indoors. Pot in general-purpose potting soil. Grow in an east or west window at 60° or less. Keep watered. Fertilize several times with liquid plant food while plant is making growth. Move outdoors into light shade in summer.

SWEET PEA *Lathyrus odoratus.* The familiar sweet pea is a vine to about 5′ grown in the vegetable garden. There are also today varieties that grow only to about 8″. Sweet pea grows wherever adequately cool weather prevails for a couple of months.

Prepare the soil in the fall. It should be dug 18″ deep and enriched with humus and bone meal. Good drainage is essential. In the spring as soon as frost is out of the ground, soak the seeds in water overnight, then plant 2″ deep. When plants are 4″ tall, sidedress with fertilizer and put in supports on which the tall vines will grow. Use twiggy brush or chicken wire. Keep

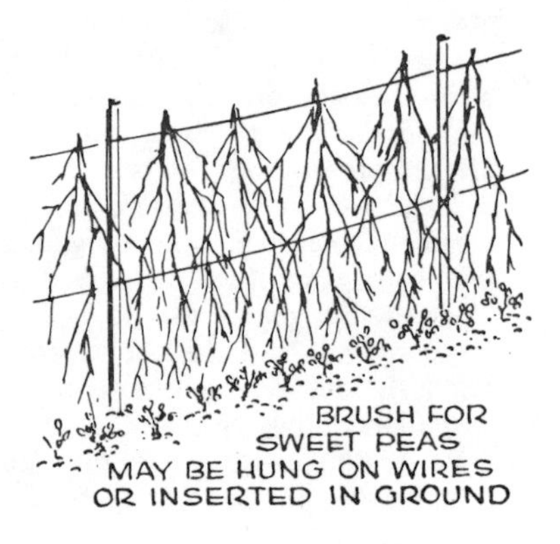
BRUSH FOR SWEET PEAS MAY BE HUNG ON WIRES OR INSERTED IN GROUND

plants well watered at all times. Make a second application of fertilizer. Cultivate carefully and frequently. Keep flowers picked.

Sweet peas are often sown outdoors in late fall just before the ground freezes. The soil is then covered with a mulch of salt hay. However, this system of planting does not necessarily produce better results than spring planting.

SWEET PEPPERBUSH *Clethra alnifolia*, spiked alder, summer sweet. Deciduous shrub to 9′ with spikes of fragrant, white or pink flowers in late summer and fall. Grows in east and southeast.

Propagate by division in the spring. Plant in spring or fall in sun or light shade in average soil that is slightly acid and holds moisture. Keep well watered. Fertilize every 2–3 years in spring.

SWEET POTATO *Ipomoea batatas* yam. Sprawling vine with slender, pointed, yellow-fleshed tubers underground. Best grown in the south because it requires 4–5 months of warm weather to mature. Also a house plant.

Outdoors. Sweet potatoes require a sandy soil enriched with a little general-purpose fertilizer with low nitrogen content. Buy slips and plant them in full sun when soil is warm. Space 20″ apart in rows 4′ apart. Weed thoroughly while vines are still small. Water in long dry spells. Lift vines from the soil when they root at the joints, but do not prune. Dig the tubers before frost. If frost strikes before it is expected, cut off the vines at once; otherwise the tubers in the ground will be ruined (but without the vines the tubers are safe and can be dug later). To cure tubers, store in a well-ventilated room for two weeks at 80°; then store in a dry place at 55°. Handle tubers with great care at all times: if bruised or cut, they rot quickly.

Indoors. Partially fill a narrow jar with water and a small lump of charcoal. Set a sweet potato tuber, fat end up, in the jar so that about one-half of the tuber is above the neck of the jar and 1″ of the bottom is in the water. Keep in a dark place until growth starts. Then move to a warm east or west window. You can continue growing the sweet potato in water; but it is better to plant it, once it is rooted, in general-purpose potting soil, which should be watered when the surface feels dry.

SWEET ROCKET *Hesperis matronalis*, dame's rocket, dame's violet, garden rocket. Biennial to 3′ with fragrant purple or white flowers. Grows in temperate climates.

Sow seeds in spring outdoors and transplant in fall. Grow in average, well-drained, humusy soil in sun or partial shade. Water and cultivate regularly. Fertilize in early spring and before flowering. The plants self-sow.

SWEETSHADE *Hymenosporum flavum*. Pyramidal evergreen to 5′ with fragrant, yellow flowers in early summer. Grows mainly in warmest parts of California.

Plant in spring or fall in average soil in sun. Water in dry weather. Fertilize in spring. To promote more open, sturdier habit, cut out some of the branches in winter.

SWEET SHRUB *Calycanthus*, strawberry shrub, Carolina allspice, spicebush. Deciduous shrubs to 12′ with fragrant, brownish spring flowers. Grows in warm climates.

Propagate by layering or by suckers. Plant in spring or fall in sun or partial shade in well-drained, average soil. Water in dry weather. Fertilize every 2–3 years.

SWEET SULTAN *Centaurea imperialis.* Annual to 4′ with fragrant flowers in various colors. Grows in cool and warm climates.

Grow like cornflower (which see), but sow seeds in the spring.

SWEET WOODRUFF *Asperula odorata.* Spreading perennial to 8″ with aromatic foliage and small white flowers. Grows in mild and warm climates.

Divide in spring and plant in light shade. The soil should contain a lot of humus and be kept moist. Fertilize in spring. Plant self-sows at will and makes a good groundcover.

SWISS CHARD *Beta vulgaris cicla,* spinach beet. A variety of beet grown as a substitute for spinach. Rhubarb chard has red leaves. Grows almost everywhere.

Plant in full sun in average soil to which some fertilizer may be applied. As soon as frost is out of the ground, sow the seeds in rows 18″ apart. Thin plants to about 6″. Water in dry weather. Start cutting leaves when about 6″ tall; new leaves will continue to appear.

SYMPHORICARPOS Different species known as coralberry, snowberry, waxberry, wolfberry, buckbrush, Indian currant. Deciduous shrubs to 7′, noted for their berries, which are usually white and long-lasting. Grows almost everywhere.

Propagate by suckers. Plant in spring or fall in partial shade or sun in average soil. Water in long dry spells. Fertilize lightly every 2–3 years. Spray with malathion to control aphids.

SYNGONIUM May be identified as *nephthytis.* Climbing or creeping plants to 2′ with philodendron-like leaves. Stems are filled with a milky juice. A house plant.

Grow like philodendron (which see).

TABLE FERN *Pteris.* Very feathery, slender ferns to 18″. Grows indoors.

For how to grow, see fern.

TAHOKA DAISY *Machaeranthera tanacetifolia,* tansy-leaf aster. Annual to 2′ with fern-like foliage and blue and yellow, daisy-like flowers. Grows almost everywhere.

Sow seeds in sun in average, well-drained soil in spring. Water in dry weather. Fertilize when thinned to stand 1′ apart.

TAMARISK *Tamarix.* Very ornamental and interesting deciduous trees and shrubs to 30′ with tiny leaves pressed close to the slender branches. Grows in mild and warm climates.

Propagate by stem cuttings. Plant in spring or fall in full sun. Tamarisk grows in almost any soil, including sand. Cut plants back hard when they are set out. Leave them alone thereafter except to prune after flowering when called for.

TANGELO *Citrus Reticulata x Paradisi.* A 25′ evergreen cross between the tangerine and grapefruit. Grows in warmest climates.

Grow like orange (which see).

TANGERINE *Citrus reticulata.* Evergreen tree to 25′ almost identical to the mandarin and satsuma. It produces flat, orange fruits with easy-peeling skins. Grows in warmest climates.

TANOAK *Lithocarpus densiflora,* tanbark oak, chestnut oak. Evergreen tree to 90′ with hairy leaves and upright flowers. Grows in California and Oregon.

Plant in spring or fall in sun in well-drained, deep fertile soil that holds moisture. Water in dry spells, especially when young. Fertilize every 3–5 years in spring.

TARRAGON *Artemisia dracunculus.* Perennial to 1′, grown for its aromatic leaves which are used as a seasoning. Grows in temperate climates.

Divide in early spring and plant in sun in average, sandy soil. Watering and fertilizing are unnecessary as a rule. Cut plants down and mulch in winter in cold climates.

TEA TREE *Leptospermum.* Evergreen trees or shrubs to 30′ with a thick covering of small leaves and white, pink or red flowers. Grows mainly in warm parts of California.

Plant in spring or fall in full sun. The soil should be of average quality, very sandy. Water and fertilize when young, but thereafter little attention of this sort is needed. Apply chelated iron to the soil according to manufacturer's directions if leaves turn yellow between the veins.

TERNSTROEMIA Handsome evergreen to 8′ with dark green, glossy foliage and fragrant, creamy flowers in spring. Grows in mild and warm climates.

Grow like *camellia* (which see), although this plant is much less fussy about soil. Grow in partial shade or sun.

TEUCRIUM CHAMAEDRYS Germander. Shrub-like perennial to 1′ with box-like foliage and rose-purple flowers. Grows in warm climates.

Propagate by cuttings in June and plant in fall or spring in a sunny location in average, well-drained soil. The crowns of the plants should be just below the soil surface. Water regularly. Fertilize in spring. Prune after flowering.

THERMOPSIS False lupine, Aaron's rod, bush pea. Perennial to 5′ with lupine-like, yellow flowers. Grows in cool climates.

Divide in fall and plant in sun in average, well-drained, sandy soil. Water regularly. Fertilize in spring with a general-purpose plant food with a low nitrogen content.

THRINAX Different species called key thatch palm, Morris Peaberry palm. Palm trees with slender trunks to 30′, crowned with fan-shaped fronds. Subtropical.

For how to grow, see palm.

THRYALLIS GLAUCA Evergreen shrub to 6′ with blue-green leaves and slender clusters of yellow flowers. Subtropical.

Sow seeds any time they are available in pots. Transplant in spring or fall in sun or partial shade. Water in dry weather. Fertilize every 2–3 years in late winter. Prune after flowering.

THUNBERGIA Different species called black-eyed Susan vine, sky flower, blue sky vine, orange clock vine, mountain creeper. Deciduous or evergreen vines to 50′ with striking flowers in various colors. Grows in warm climates.

Propagate by stem cuttings in the spring. (It is best to start new plants of black-eyed Susan vine by this means—or by seeds—every spring.) Plant in good soil with extra humus. Grow in sun but protect from wind. Fertilize in spring and again in summer. Water in dry weather. Tie branches to a lattice. Prune in early spring. Spray with malathion to control red spider.

THYME *Thymus.* Perennial herb to 8″, the small leaves of which are used for flavoring. Grows almost everywhere. See also mother of thyme.

Sow seeds in spring in a sunny spot in average soil which is very well drained. Keep moist until plants are established; thereafter, water in dry spells. Fertilize in spring. Protect with salt hay or evergreen boughs in winter. To dry for storage, hang upside down in a warm, airy, shady place.

TIGRIDIA Mexican shellflower, Mexican tiger flower. Bulbous plants to 30″ with cup-shaped, red flowers with brightly spotted centers. Grows almost everywhere.

Grow like *gladiolus* (which see). Plant bulbs 4″ deep and 6″ apart.

TILLANDSIA Mainly ephiphytic bromeliads to 10′ with violet-blue flowers. Some have spreading leaves forming a vase-like center. Some have leaves covered with scales. Grows in subtropics; also indoors.

Pot in osmunda fiber in well-drained pots. Grow in an east or west window or outdoors in filtered sun. Keep reservoirs of those which have such filled with water at all times. Spray plants covered with scales with water every day or two, and keep them in a place where there is good air circulation. Fertilize all species with dilute liquid plant food every six weeks while plants are making growth. Protect outdoor plants from frost. Spray with malathion if infested with scale insects, but rinse the next day. Species which produce offsets can be propagated by planting these in osmunda fiber.

TITHONIA Mexican sunflower. Perennial to 8′ grown as an annual in mild and warm climates, with dahlia-like, orange-yellow or orange-red flowers in late summer.

In warm climates, sow seeds outdoors where plants are to grow in spring. In cooler climates, start seeds indoors 8–10 weeks before last frost. Grow in full sun in average soil. Fertilize when plants are established. Water only in dry weather.

TOMATO *Lycopersicum.* Favorite vegetable to 6′. Grows almost everywhere.

Buy plants or sow seeds indoors 8–10 weeks before last frost. When seedlings develop first true leaves, transplant into larger flats or individual small pots. Grow in a sunny window. Keep soil barely damp. Ten days before plants are to go into the garden, stop watering entirely and move them outdoors every day (but not night) to harden off.

Plant outdoors in full sun when all danger of frost is past. Prepare average, well-drained soil by digging deeply and mixing in several spadefuls of humus. Space planting holes 2′ apart in rows 3′ apart. Set the plants at the depth they formerly grew or deeper (roots grow from the stems). Water thoroughly with a transplanting hormone or liquid plant food. Thereafter, fertilize monthly either with liquid plant food or dry general-purpose fertilizer. Keep well watered and weeded. Mulch the soil to hold in moisture. Dust plants about every ten days with a combination insecticide-fungicide.

Tomatoes can be allowed to sprawl over the ground but they produce somewhat better fruit if trained to stout stakes at least 4′ tall. Nip out all but about one-quarter of the side branches. Before first frost, pull up plants and hang them in a protected place. Many of the green tomatoes on the vines will ripen.

TORENIA Annual with dainty, snapdragon-like flowers in blue and yellow. Grows almost everywhere.

Sow seeds indoors about eight weeks before last frost, then transplant into average soil in partial shade. Fertilize when young plants are established. Water regularly.

TORREYA Different species called California nutmeg and stinking cedar. Evergreen trees to 75′ with spreading branches and yew-like foliage. Most species grow only in warm climates, but one survives considerable cold.

Plant in spring or early fall in sheltered, partially shaded spot. The soil should be of average quality, well drained but moisture-retentive. Water in long dry spells. Fertilize every year or two while trees are young.

TRAILING ARBUTUS *Epigaea repens*, Mayflower. Spreading, evergreen shrub only a few inches tall with fragrant, white or pinkish flowers in early spring. Grows in cold and cool climates.

Buy plants or dig from the wild with as much soil and root as possible. Plant in spring in partial shade in acid soil containing considerable sand and some humus. Mulch annually with pine needles or oak leaves. Water in dry spells. This plant is difficult to grow.

TRICHOCEREUS Organ pipe cactus. One species known as white torch cactus. Spiny desert cacti to 70′ with stems rising like organ pipes from the ground. Grows in warm climates where temperatures do not fall below 20°. Also grows indoors.

For how to grow, see cactus, desert.

TRILLIUM Wake robin, American wood lily, ground lily, trinity lily, Jew's harp, birthroot, nosebleed, bloody butchers. Perennials to 18″ with three leaves arranged in a whorl and flowers in various colors with three petals. One species or another grows in almost every part of the country.

Sow seeds right after they are harvested, or divide roots in the fall. Plant in spring or fall in shade. The soil must be moist and very humusy Water in dry weather. Fertilize lightly in very early spring.

TRIPTERYGIUM REGELI Deciduous shrub to 6′ with reddish-brown stems and clusters of fragrant, greenish-white flowers in summer.

Grows in all but coldest climates.

Grow from seeds sown in spring in average soil that retains moisture. Needs sun. Water in dry weather. Fertilize and prune in early spring. Tie stems to a trellis, wall or post for a neater appearance, because stems are straggly.

TRITONIA Blazing star. Often identified as *montbretia.* Bulbous plants to 3′ with long spikes of tubular, orange or red flowers in summer. Grows almost everywhere.

Grow like *gladiolus* (which see). Plant bulbs 4″ deep and 5″ apart in partial shade. Don't let bulbs in storage dry out—occasionally dampen the peatmoss in which they are stored. In warm climates bulbs may be left in the ground over winter.

TROLLIUS Globeflower. Perennials to 2′ with ball-shaped, yellow or orange flowers. Grows in temperate climates.

Divide in spring or fall and plant in partial shade in average soil that is humusy and always moist. (The plants do very well beside ponds, streams, etc.) Fertilize in spring. Pick off seed pods.

TRUMPET CREEPER *Campsis,* trumpet vine. Deciduous vines to 30′ with clusters of trumpet-shaped flowers in red. Grows in all but coldest climates.

Propagate by layering. Plant in sun in spring or fall in good, well-drained loam improved with humus and fertilizer. Keep moist and fertilize once or twice during the spring and summer with superphosphate. Although the vine puts out aerial rootlets which cling to rough surfaces, provide additional support to ensure that it will not pull loose in heavy winds or of its own weight. Cut back the branches in early spring to about two buds. Pinch back stem ends occasionally to promote bushier growth. Spray with malathion to control aphids.

TUBEROSE *Polianthes tuberosa.* Tuberous plant to 42″ with very fragrant, waxy, white flowers in summer. Grows almost everywhere.

Plant the tubers 2″–3″ deep and 6″ apart in well-drained, rich, humusy soil when all danger of frost is past. Grow in the sun. Keep watered. Fertilize when growth appears. Dig up tubers in the fall and store in peatmoss or vermiculite at normal house temperatures. Don't count on bloom two years in a row.

TUBEROUS BEGONIA Upright or pendulous, tuberous-rooted plants with extremely beautiful flowers in reds, pinks, white, yellows, oranges. Grows best in the fog belts along the Pacific and New England coasts, but may flower in other temperate regions in summer and fall. See also *begonia.*

For early bloom, start tubers in pots indoors about eight weeks before last frost. Or plant tubers outdoors (in 8″ pots or directly in the garden) when the weather is reliably warm. Tuberous *begonias* need partial shade.

Plant the tubers round side down in well-drained soil containing considerable humus and a little dried manure. Cover the tiny pink buds on the concave side of the tubers with ½″ of soil. Water well and keep the soil evenly moist, but not soggy. As long as the foliage is green, apply liquid plant food once every 3–4 weeks. If foliage turns blue-green, stop feeding until it turns soft green again. But if foliage turns pale green, increase feeding to every two weeks.

Dust with sulfur occasionally to prevent mildew. Other apparent ills are probably not attributable to insects or diseases. For example, if a plant has good foliage but no flowers, you may be shading it too deeply. If a plant is stunted and the leaves are thick and satiny, you may be giving it too much sun. If buds and flowers drop, you may be giving the plant too much or too little water, or the weather may be too hot.

In the fall, when the foliage dies down, dig up the tubers and let dry. Leave the foliage alone until it can be removed with a gentle tug. Remove all particles of the stem. Store the tubers in trays in a dry, cool (40°–50°) place. If they show signs of shriveling, place them in barely damp peatmoss.

TULIP *Tulipa.* Favorite spring bulbs, from 4″ to 3′, with flowers in practically all colors. Grows everywhere. Also indoors.

In most parts of the country. Plant bulbs in a sunny spot any time in the fall until the ground freezes hard. The soil should be dug 12″–15″ deep and enriched with a little bone meal. If it is heavy, mix in humus and sand. If drainage is not good, place a 2″ or thicker layer of sand at the bottom of the hole. Tulips need plenty of moisture but must have good drainage.

Cover the bulbs of small, early-blooming species with 5″ of soil; larger types with 9″–12″ of soil. Make the holes with a trowel so that they are flat on the bottom. Space bulbs 6″ apart. If your garden is a haven for rodents, enclose the bulbs in baskets made of ½″ wire mesh or (less good) dust them with anti-rodent powder. Water newly planted bulbs well; thereafter water only in dry weather.

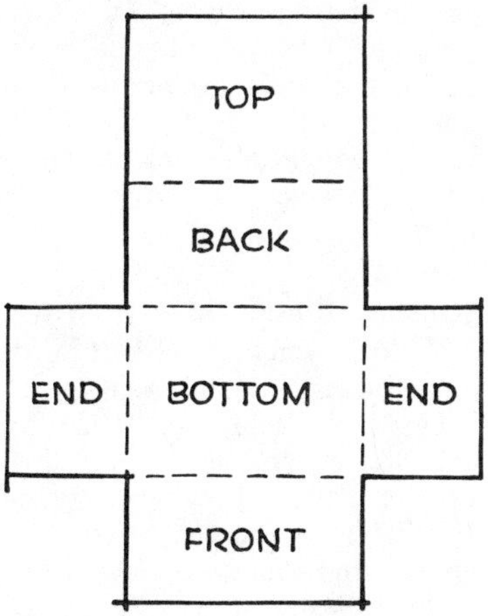

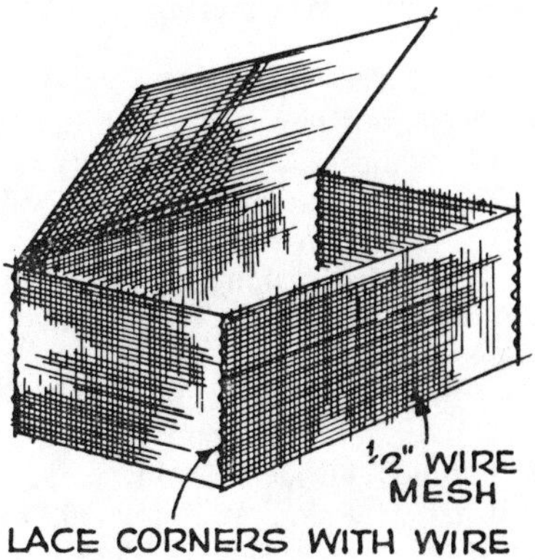

In the spring, after flowers have bloomed, snap off the stems just under the flower heads, but don't touch the foliage until it has withered and can be pulled from the bulbs with a slight tug. In the fall, scratch a little general-purpose fertilizer into the soil.

In warm climates. In areas where the ground does not freeze, plant special pre-cooled bulbs. Or pre-cool your own bulbs by storing them in the bottom of the refrigerator for six weeks. Don't let the bulbs freeze. Then plant the bulbs no more than 4″–5″ deep, preferably in a spot that is shady in winter.

In the Rockies. Plant tulips in partially shaded spots. Mulch the soil heavily after it freezes in the fall.

Indoors. See How to Force Bulbs for Indoor Bloom.

TULIP TREE *Liriodendron tulipifera*, yellow poplar, whitewood. Deciduous tree to 150′ with greenish-yellow, tulip-shaped flowers in early summer. Grows mainly east of the Mississippi.

Plant in early spring before growth starts. The specimen should be under three years old and should have been well root-pruned a year prior to transplanting. Grow in sun in rich, moist soil. Fertilize every 3–5 years. Water in dry spells.

TUNG OIL TREE *Aleurites fordi*. Deciduous tree to 20′ with clusters of white and red flowers and poisonous fruits from which valuable oil is pressed. Grows best in a narrow band along the Gulf Coast and across Florida and Georgia to the Atlantic.

Plant in spring or fall in sun. The soil should be of good quality, containing considerable loam; somewhat acid and well drained. Fertilize every spring. Water in dry weather.

TUPELO *Nyssa*, sour gum, pepperidge. Deciduous tree to 100′ with spectacular fall foliage. Grows in cool and warm climates.

Plant young specimens only. They should have been well root-pruned in previous years. Plant in early spring or fall in sun in moist or even wet, average soil. Water copiously until established; thereafter in all dry weather. Fertilize every 3–5 years.

TURFING DAISY *Matricaria tchihatchewi*. Spreading perennial with fern-like foliage and tiny daisies on 6″ stems. Used as a groundcover. Grows almost everywhere.

Divide in spring and plant in sun in almost any soil. Water until plants take hold, then only when they look in poor shape.

TURK'S CAP *Malvaviscus*, sleeping hibiscus. Shrubs to 10′ with more or less heart-shaped leaves and red or pink flowers much of the year. Grows in warmest climates.

Propagate by greenwood stem cuttings. Plant in spring or fall in sun or partial shade. Turk's cap grows in average soil but prefers one that is light, well-drained and humusy. Water in dry spells. Fertilize a couple of times during the year. Plant should revive if killed back by frost.

TURNIP *Brassica rapa*. Root vegetable to 2′ which grows best in cool climates. Also see rutabaga.

Dig soil deeply, pulverize well and add general-purpose fertilizer, rich in potash. Sow seeds in late July or early August in rows 15″ apart. Thin plants to 3″–4″. Keep weeded. Water in dry weather. Dig before ground freezes. Seeds of several varieties of turnips may also be sown in the spring as soon as the ground can be worked. They are harvested before the weather turns warm.

TURTLEHEAD *Chelone.* Perennials to 2′ with white or reddish-purple flowers in spikes. Grows in the east and southeast.

Divide plants in spring and plant in partial shade in average, moist soil. Fertilize in spring.

TWINFLOWER *Linnaea.* Trailing, evergreen vine only a few inches tall with tiny pink flowers. Grows in cold and mild climates.

Propagate by stem cuttings in spring. Plant in spring or fall in acid, moist, well-drained, humusy soil. Grow as a groundcover. Fertilize if plants are not growing well.

UMBRELLA PINE *Sciadopitys verticillata.* Handsome conifer to 100′ with long needles grouped around the shoots like the ribs of an umbrella. Grows in mild and warm climates.

Plant in spring or early fall in a lightly shaded location protected from the wind. Grow in average, moisture-retentive soil. Water in drought. Fertilize every 3–5 years.

UMBRELLA PLANT *Cyperus alternifolius.* A bog plant with several thin, 4′ stems topped with umbrella-shaped leaf clusters. Mainly a house plant.

Propagate by stem cuttings in water or moist sand. Pot in general-purpose potting soil. Grow in a south window. Keep soil constantly wet at all times by standing the pot in a deep saucer of water. Fertilize occasionally.

UMBRELLA TREE *Schefflera actinophylla.* Evergreen tree to 25′ with umbrella-shaped leaf clusters and red flower spikes. Mainly a house plant, but also grows outdoors in subtropics.

Indoors. Pot in general-purpose potting soil. Grow in an east or west window. Fertilize occasionally. Water when soil feels dry. Move into partial shade outdoors in summer. Air-layer tree when it grows too large for the house.

Outdoors. Plant in partial sun in average soil. Water in dry weather. Fertilize in spring. Spray twice in the spring with an oil emulsion spray to control scale.

UNICORN PLANT *Proboscidea jussieui,* devil's claw, proboscis flower. Sprawling plant with stems about 3′ long, funnel-shaped flowers and large, curved, hanging seedpods used for pickling. A perennial in the south; grows as a tender annual in cool climates.

Needs a light, humusy, well-drained, limed soil. In the south sow seeds outdoors in the spring where the plants are to grow. In the north, sow seeds indoors 8–10 weeks before last frost. Grow in the sun everywhere. Water in long dry spells. Fertilize when young plants are established. Keep weeded.

VALERIANA Valerian, garden heliotrope, cherry pie, cretan spikenard. Perennials to 5′ with rounded clusters of fragrant, pink, white or purple flowers in summer. Grows in temperate climates.

Divide plants in spring or fall and plant in a sunny spot. Grow in average soil which should be limed every year or two and fertilized in the spring. Keep fairly moist and cultivated. The plants self-seed.

VANDA Epiphytic orchids, some climbing to 10′, with large clusters of flowers in many colors at different times of year. A house plant.

For how to grow, see orchid. *Vanda* needs considerable humidity but also good ventilation (without exposure to chilly drafts). Shading from midday sun is necessary only in the south.

VANILLA TRUMPET VINE *Distictis lactiflora.* Evergreen vine to 30′ with leathery leaves and sweet-scented, purple, summer and fall flowers. Grows in warmest parts of California.

Propagate by stem cuttings or seeds. Plant in well-drained, average soil in partial shade and away from strong winds. Keep soil moist. Fertilize in spring. Provide support that tendrils can wrap around. Thin out the stems in the fall.

VEITCHIA Palm trees with single trunks up to 100′ and with a spreading crown of more or less featherlike fronds and red fruits. Subtropical.

For how to grow, see palm.

VELTHEIMIA Bulbous house plant to 30″ with a pointed cluster of rose-pink flowers in winter.

Pot in a 6″ container in general-purpose potting soil so that bulb is half exposed. Do not over-water until growth starts, then apply water whenever soil feels dry. Place pot at first in north window, then move to a south window, then to an east window when flowering starts. Fertilize twice during the three-month growing period. When flowering stops, cut off flower stalk and gradually reduce water. When foliage dies, stop watering entirely and store the bulb in its pot in a dark, cool place.

VELVET PLANT *Gynura aurantiaca.* Shrub to 4′ with foliage and stems covered with purple hairs, and sometimes with orange flowers. A house plant; also grows outdoors in subtropics.

Indoors. Propagate by stem cuttings in summer. Pot in general-purpose potting soil with a little extra humus or well-rotted manure. Grow in a warm south window. Keep soil moist. Spray foliage often with water. Fertilize lightly 3–4 times a year.

Outdoors. Plant in partial shade in humusy soil. Water in dry weather. Fertilize in spring and fall. Prune to control growth and discourage self-layering.

VENIDIUM Annual to 3′ with orange flowers. Grows best in warm climates.

Sow seeds indoors about eight weeks before last frost, or sow outdoors where plants are to grow when weather is warm. *Venidium* does best in average soil with just a little fertilizing. Water in very dry weather. Full sun is needed.

VENUS FLYTRAP *Dionaea muscipula.* Perennial with a low rosette of leaves which trap insects and with white flowers on a 1′ stalk. Grows in the Carolinas; also indoors.

For how to grow, see pitcher

plant. Keep house plants at about 60° in a humid atmosphere. Remove flower stalks as they develop.

VERBENA Perennial plants to 5′ with large blue, purple, white or rose flower trusses in midsummer. Grown almost everywhere as annuals.

Start seeds indoors 8 to 10 weeks before last frost. Transplant into individual pots. When weather is reliably warm, plant in full sun in average soil. Fertilize lightly when plants are established. Water regularly. Keep seed pods picked off.

VERONICA Speedwell. Perennials to 2′ with mainly blue flower spikes. Grows in all but extreme climates.

Divide plants in spring or fall and plant in average, well-drained soil. Give sun or partial shade. Do not let plants stay dry too long. Fertilize in spring. Cut off seed pods after flowering.

VIBURNUM Different species called snowball, dockmackie, hobblebush, witch-hobble, witherod, arrowwood, wayfaring tree, nannyberry, sheepberry, poison haw, cranberry tree, black haw, stag bush. Large genus of shrubs or trees to 30′, mainly deciduous, with lovely white, pink or lilac, fragrant flower clusters followed by red, black or blue fruits. Depending on the species, grows almost everywhere.

Propagate by layering. Plant in spring or fall. All viburnums do well in the sun and some also do well in light shade. Grow in average soil. Water in dry weather. Fertilize in early spring. Prune early-blooming varieties after flowering; others in late winter.

VINE LILAC *Hardenbergia.* Evergreen vine to 8′ with dripping, lilac-blue flower clusters in early spring. Grows best in California.

Propagate by young stem cuttings in spring. Plant in well-drained light soil which should not be kept too moist. Grow in partial shade. Fertilize before and after flowering. Train branches on a sturdy support. After flowering, cut back flowering branches and long side branches.

VIOLA Violet. Favorite spring- and summer-flowering perennials to 1′. Grows best in temperate climates.

Propagated by seeds sown in early spring or by divisions in early spring or fall. Transplant into partially shaded locations. Violas generally prefer a humusy, moist soil enriched with a little general-purpose plant food in the spring, but a few wild violets grow in dry, rather barren soil. Water during hot spells. Spray with an all-purpose insecticide-fungicide as necessary. Keep dead flowers picked off.

VIRGINIA CREEPER *Parthenocissus quinquefolia,* woodbine, American ivy. One of several species of handsome, deciduous vines to 40′ with blue-black berries in the fall. Grows in all but very coldest regions.

Propagate by layering or stem cuttings. Plant in average soil. Fertilize lightly in spring. Water in very dry weather. The vine grows in sun or light shade. The root-like holdfasts cling to any rough surface. Prune in early spring just enough to control direction of growth.

VIRGINIA STOCK *Malcomia maritima,* Mahon stock, Malcolm stock.

Annual to 1′ with small pink, red, white or purple flowers. Grows almost everywhere.

Sow seeds in early spring where plants are to grow and thin to stand 8″ apart. Grow in sun in average soil. Water in dry weather. Fertilize lightly when plants are thinned. Will probably self-sow.

VITEX Chaste tree, hemp tree, monks' pepper tree. Deciduous shrubs to 12′ with blue or white flower spikes in summer or early fall. Grows best in hot, dry climates.

Propagate by cuttings. Plant in spring or fall in a sunny location in average soil. Water sparingly in dry weather and fertilize every 2–3 years in spring. If plants are killed back by cold, they usually put out new growth from the base. Prune in early spring.

VRIESIA One species known as flaming sword. Mainly epiphytic bromeliads to 3′ with leaves forming cups or reservoirs and with long-lasting flowers in various colors, but usually white or yellow. Grows in subtropics; also indoors.

Propagate by offsets planted shallowly in equal parts of sand and peatmoss. When rooted, pot up in osmunda fiber in well-drained pots. Grow in a north window or in shade outdoors. Water when fiber feels dry. Keep reservoir in center of leaves filled with water at all times. Spray house plants with water frequently, Give a little dilute liquid plant food every six weeks while plant is growing. Protect outdoor plants from frost. Spray with malathion to control scale if it appears, but rinse plants with clear water the next day.

WALKING FERN *Camptosorus.* Evergreen fern with undivided, 12″ leaves that take root at the tips. Grows in temperate climates east of the Rockies.

For how to grow, see fern. Propagate by burying leaf tips (although plants will do this by themselves). Grow in light shade. Apply lime to soil if it is acid.

WALLFLOWER Name applied to flowers in the genus *cheiranthus* and also to those in the genus *erysimum.* Annuals and perennials to about 30″ with spring or summer flowers in terminal clusters in various colors. Grows best in temperate climates.

Though somewhat different in character, these wallflowers are grown in basically the same way. Start annual seeds outdoors in the spring where plants are to grow. Grow perennials from spring-sown seeds or from divisions made in spring. Grow in full sun in average soil which must be very well drained. Fertilize in spring. Water in long dry spells.

WALNUT *Juglans.* One species known as butternut. Handsome deciduous trees to 150′ with feathery foliage and large edible nuts. One species or another grows almost everywhere.

Plant in spring or fall in full sun. As a group the walnuts require only deep, average, well-drained soil; but the English walnut requires very fertile, medium-weight soil without any trace of alkalinity. Prune young trees annually in winter for the first couple of years to develop a straight, well-shaped, tree-like structure (when young, walnuts tend to grow every which way). Water deeply in dry weather. Fertilize in spring for the first 2–3 years; thereafter every 3–5 years.

WANDERING JEW Name given to *Tradescantia fluminensis*, which has striped or pale leaves and white flowers, and to *Zebrina pendula*, which has reddish-purple leaves and purple flowers. Both are trailing plants with stems to about 3′. Mainly house plants, but also grow outdoors in the subtropics.

Indoors. Propagate by stem cuttings in water or moist sand. Pot in general-purpose potting soil. Grow in a north window. Keep moist at all times. Fertilize every two months. Wandering Jew also grows in a jar of water.

Outdoors. Use as a groundcover. Plant in shade in average soil. Water in dry weather. Plants spread fast.

WASHINGTON PALM *Washingtonia*. Palm trees to 90′ with fan-shaped fronds. The old leaves hang down around the trunks below the live foliage in thick skirts. Grows in warmest climates.

For how to grow, see palm.

WATER ARUM *Calla palustris*, bog arum, wild calla. Perennials to 9″ with shining, heart-shaped leaves, white, calla-lily-like flowers and red berries. Grows mainly in cold climates.

Take seeds from berries when ripe and sow in silty loam in flats under water. Transplant into permanent positions in the fall. Plants need full sun and an acid, wet soil that may be under as much as 2″ of water.

WATER CRESS *Nasturtium officinale*. Small, pungent, perennial salad plant growing in running water in cool weather.

Propagate by dividing plants in spring or fall. Set in humusy soil in water that just covers. Cut before seed stalks form.

WATER LILY The water lilies belong to several genera. Some are hardy; some tropical. All can be grown anywhere.

Hardy lilies. Propagate by cutting the fleshy roots into about 6″ lengths, each containing one or more buds or growing points. Plant in the spring shortly before the last killing frost. The roots should be kept moist while out of the ground.

Water lilies are best planted in containers about 10″ deep and 14″–18″ wide. Use any good loam (but not pond muck) and mix in dried manure at the rate of about one part manure to ten parts soil. Plant the roots of tuberosa and odorata water lilies horizontally and 1″ deep. The growing points should extend above the soil surface. Marliac roots are planted upright with the crown above the soil surface. If there are

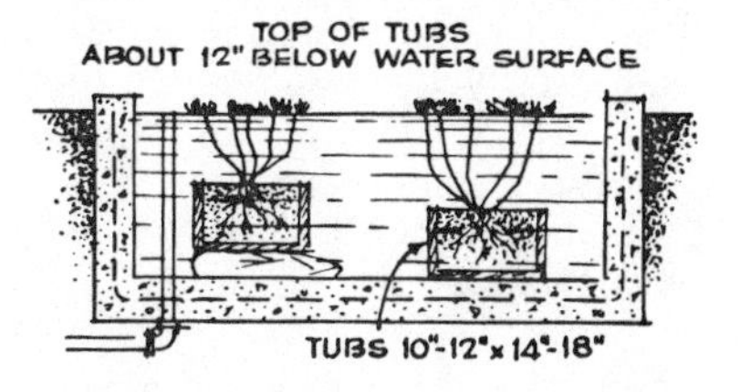

fish in your pool, cover the soil with pebbles so that the fish will not roil the water. Be careful always not to injure the growing points of the tubers.

Let the water in the pool warm

up for several days before placing the containers in it. The sudden shock of very cold water can set the plants back badly. Cover the plants with only 1″–2″ of water so that the sun will stimulate growth of the growing points. Then when growth starts in about a week, drop the container to the bottom of the pool or cover with more water. On the average, water lilies do well when covered with about 1′ of water, but some are happy with less and others with considerably more. Full exposure to sun is required. Allow about 10 sq. ft. of water surface per plant.

Additional fertilizer is not necessary the first year but should be applied every spring thereafter. Use a general-purpose plant food and work a couple of handfuls of it carefully into the soil. Fertilizer may also be needed if the plant does not appear to be growing well during the summer. Repot the plants in fresh soil every third spring. The roots can be trimmed back hard and divided at this time.

If the containers are below the frost line, the plants can be left underwater in winter without any special protection. Otherwise, cover the pool with boards and leaves or straw. Or you can dig up the roots and store them in a cool place indoors in moist sand.

During summer, keep dead blooms picked off.

Tropical lilies. Handle as above with the following exceptions: Plant in late spring when water temperature is about 68° or higher. When growing, the plants need only about 8″ of water over them. Fertilize once or twice during the summer by poking plant tablets into the soil around the roots.

Leave propagation of tropical lilies to experts. Since the plants cannot go through the winter under water except in warmest climates, they must be dug up and stored. But the chances are that they will then rot. Therefore it is better to buy new roots every spring from a reliable dealer.

WATERMELON *Citrullus vulgaris.* Spreading, annual vine best grown in warm climates. However, new midget varieties are available for the north, where they do very well.

The soil must be light, sandy, well-drained and improved with general-purpose fertilizer. Full sun is required. After all danger of frost is past and the soil is warm, plant 4–8 seeds 1″ deep in groups at least 6′ apart in all directions. Later, thin to one strong plant. Keep dusted with rotenone until flowering begins. Pull out weeds. Water regularly and well. Fertilize once or twice after thinning plants. Pick watermelons when they give off a dull, muffled sound when you rap them with your knuckles.

WATSONIA Bugle lily. Plants growing from corms to 4′ with summer flowers much like *gladiolus* in white, pink or red. Usually grown in warm California and the south, but also grows in cooler climates.

Grow like *gladiolus* (which see). In the deep south, the corms are planted in the fall and can be left in the ground through the years until they become crowded.

WAX PLANT *Hoya carnosa.* Evergreen vine to 10′ with shiny green foliage and fragrant, waxy, white or pinkish flowers in spring and summer. Best known as a house plant. Also grows in subtropics.

Indoors. Propagate by layering or by tip cuttings in spring. Pot in general-purpose potting soil. In fall and winter, water very sparingly and keep the plant in a cool, not too light place. From February on to the end of flowering, keep in a warm south window. Water thoroughly. Fertilize about every three weeks. Spray foliage with water. Provide a rough support that the aerial rootlets can cling to. Prune the plant very sparingly in early spring if it seems necessary, but do not remove the short spurs which produce flowers.

Outdoors. Plant in average soil with extra humus. Grow in partial shade. Protect from wind (air circulation must, however, be good). Spray with malathion to control mealybugs. Otherwise, follow directions above.

WEIGELA Deciduous shrub to 10′ with handsome clusters of funnel-shaped flowers in various shades of red from about May through July. Grows in all but coldest climates.

Propagate by stem cuttings in summer. Plant in spring or fall in full sun in average soil that holds moisture. Water in dry weather. Fertilize lightly in early spring. After flowering, cut worn-out stems to the ground.

WILD CUCUMBER *Echinocystis lobata.* Fast-growing, annual vine to 20′ with white flowers and puffy fruits. Can become a pest, but useful for concealing unsightly objects. Grows almost everywhere.

Sow seeds in the late fall where vine is to grow. Grow in almost any soil, sun or shade. The plant needs little care; self-sows badly.

WILD GINGER *Asarum.* Creeping perennial to 10″ with heart-shaped leaves, insignificant flowers and aromatic roots. Grows in cool climates.

Divide plants in spring and plant in shade in rich, moist, very humusy soil. Water in dry weather. If the plant does not spread rapidly, fertilize in early spring.

WILD STRAWBERRY *Fragaria.* Perennials to about 1′, the antecedents of the cultivated strawberries. Often grown as groundcovers. Different species grow in different parts of the country. See also strawberry.

Propagate by the rooted runners. Plant in fall or early spring in sun in average, well-drained, humusy soil. Water regularly. Fertilize in early spring if the plants need a boost.

WILLOW *Salix.* Deciduous trees to 60′ with soft, brittle wood and narrow, bright-green leaves. The handsomest species are the weeping willows. Grows almost everywhere.

Plant in spring or fall in a sunny location. The trees do best in moist soil of average quality, but also get along in drier locations. Don't plant near septic fields, because the roots grow straight for the pipes. An application of fertilizer every 3–4 years helps plants but is not essential. Prune out dead and broken wood regularly. Plants are easily propagated by stem cuttings in the spring.

WINGED EVERLASTING *Ammobium.* Annual everlasting to 3′ with winged branches and silvery-white bracts. Grows almost everywhere.

Sow seeds outdoors when danger of frost is past. The plants need sun, average soil. Water in dry weather. Fertilize when plants are thinned to 9″. Cut flowers before fully open and hang them upside down in a shady place to dry.

WINTER ACONITE *Eranthis.* Tuberous-rooted plants to 8″ with yellow, buttercup-like flowers in late winter. Grows best in cool climates.

Plant tubers as soon as you receive them, but if they look shriveled, put them in moist peatmoss or sand overnight. Plant 2″–3″ deep and about 3″ apart in moist, humusy soil in light shade.

WINTERSWEET *Chimonanthus praecox.* Deciduous shrub to 10′ with fragrant, yellow flowers in winter. Grows in warm climates.

Propagate by layering. Plant in a sunny, sheltered location in spring or fall. Grow in average soil. Water in dry weather. Fertilize every 2–3 years in fall. Prune after flowering.

WISTERIA Deciduous vine to 60′ with light foliage and magnificent, dripping clusters of violet or white flowers in spring. Grows in all but coldest climates.

Plant grafted plants in spring or fall in good soil that has been deeply dug and reinforced with humus. Apply a little general-purpose fertilizer every 2–3 years in early spring. Keep soil moist in summer. Full sun is required. Grow on a very sturdy trellis and guide branches, otherwise they will become hopelessly tangled. If the vine is grown near a house, watch out for tendrils that creep under shingles and clapboards.

If flowering is good, prune only enough to keep vine neat and under control. If flowering is inadequate, cut back long stems (but not the main stems) about half in summer. This should encourage flowering but sometimes it does not. In the latter case, cut roots 2′ from the main stem in October. Note, however, that it is unwise to undertake any drastic pruning until you have let plant go its own way for several years.

WITCH HAZEL *Hamamelis.* Deciduous shrubs to 25′ with mostly yellow, fragrant flowers in fall, winter or early spring. Grows in mild and warm climates.

Plant young specimens only in the spring. Grow in partial shade in average soil which holds moisture. Water in dry weather. Prune after flowering. Fertilize every 3–4 years.

WOODSIA Delicate, little ferns (under 1′) growing in cool climates.

For how to grow, see fern. These ferns grow naturally in shaded pockets in rocks. The soil must be cool and moist.

XANTHOSOMA Tuberous plant to 3′ with large, handsome, arrow-shaped leaves. Grows indoors.

Divide tubers in summer, making sure each has an eye. Pot in two parts loam, two parts humus and one part sand in a tight-fitting pot. Grow in a warm east or west window. Keep soil moist at all times. Spray foliage with water often. Fertilize 3–4 times a year.

XERANTHEMUM ANNUUM *Immortelle.* Annual everlasting to 3′ with white or purple flowers and papery bracts. Grows almost everywhere.

Sow seeds in the spring in average

soil. Grow in full sun. Water in dry weather. Fertilize when young plants are thinned to 9″. Hang flowers upside down in a shady place to dry.

XYLOSMA SENTICOSA Easily trained, evergreen shrub to 6′ with foliage that is reddish-bronze in spring, green later. Grows best in hot, dry climates.

Plant in average soil in sun in spring or fall. Fertilize every 1–2 years in early spring. Water in very dry weather. Train branches to a lattice or on a wall or let them spill over a wall. Prune in early spring to control shape and eliminate dead wood.

YARROW. *Achillea*, milfoil, sneezewort. Perennial to 2′ with aromatic foliage and flat clusters of pink, white or yellow flowers. Grows almost everywhere.

Propagate by spring divisions, or sow seeds indoors about eight weeks before last frost. Give plants full sun and an average soil which is fertilized in the spring. Water in dry weather.

YELLOW ELDER *Stenolobium stans.* Also identified as *Tecoma stans.* Deciduous shrub to 15′ with yellowish green foliage and fragrant, red-striped, yellow flowers in fall and winter. Grows in warmest climates.

Propagate by self-sown seedlings. Plant in spring or fall in sun in average soil. Water in dry weather. Fertilize in summer. Prune after flowering stops in late winter or spring. Protect from frost in cold areas.

YELLOW OLEANDER *Thevetia nereifolia*, lucky nut. Evergreen shrub or tree to 8′ with dense foliage; yellow, trumpet-shaped flowers, and black fruits. Very poisonous. Grows in warmest climates.

Propagate by semi-hardwood stem cuttings. Plant in sun in average soil containing extra sand. Water when soil feels dry. Fertilize annually. Prune after flowering to maintain shrubby growth and prevent plant from developing into a tree.

YELLOW ROOT *Zanthorhiza simplicissima.* Deciduous shrub to 20′ with clusters of brownish-purple flowers in spring and bright-yellow foliage in fall. Grows in all but coldest parts of the east.

Propagate by division in spring or fall. Plant in partial shade in average, moist soil. Water in dry weather. Fertilize every 2–3 years.

YELLOW STARGRASS *Hypoxis hirsuta.* Bulbous plant to 9″ with grass-like leaves and little, yellow, star-shaped flowers. Grows mainly in the east.

Plant bulbs in the fall 1″ deep and 4″ apart in filtered sun in average, sandy soil. Give a light dose of fertilizer every 2–3 years. Dig up and divide when plants get too thick.

YELLOW WOOD *Cladrastis lutea*, gopherwood. Densely foliaged, rounded, deciduous tree to 50′ with wisteria-like clusters of fragrant, white flowers in late spring and early summer. Grows in mild climates.

Plant in spring or fall in a sunny location. Can grow in average soil but does best in deep, rich soil. Water in dry weather. Fertilize every 2–4 years in the spring. Prune in the fall if necessary.

YEW *Taxus*. Evergreens ranging from smallish shrubs to 40′ trees, all with beautiful, wide, dark-green needles. Foliage and fruit are poisonous. Grows in all but extreme climates, though one species known as ground hemlock (*T. canadensis*) is hardy far north.

Plant in spring or early fall in average soil to which extra humus has been added. Grow in sun or partial shade. Water in dry weather. Feed every 3–4 years. Most varieties need to be pruned annually in the spring. They may also be sheared like a hedge. Note that deer have an insatiable appetite for yews.

YOUNGBERRY A bramble fruit—a form of dewberry which grows in warm climates.

Grow like blackberry (which see). Canes should be trained to horizontal wires.

YUCCA Different species called Spanish bayonet, Spanish dagger, Joshua tree, Adam's needle. Plants with tough, evergreen, sword-shaped leaves in rosettes and beautiful clusters of white flowers. Most species have no stem but send up tall flower stalks. Several species are trees with trunks to 40′. Grows mainly in the southwest, but different species are to be found almost everywhere except in coldest climates.

Smaller species may be propagated by division of the roots. Plant in spring in full sun in light, sandy or gravelly soil which is well drained. Once established, yucca can survive almost any abuse.

ZANTHOXYLUM Also identified as *xanthoxylum*. Different species called prickly ash, angelica tree, toothache tree, Hercules' club. Deciduous, prickly, aromatic trees or shrubs to 50′. Grows in east and south.

Plant in early spring or fall in sun in average, well-drained soil. Water in very dry weather. Fertilize every 3–5 years.

ZAUSCHNERIA Sprawling perennials to 30″ with grey-green foliage and red flowers in late summer. Grows in California.

Divide and plant in spring or fall in full sun in average soil. Water until established; thereafter, plants can pretty well take care of themselves. Fertilize lightly in spring. Pinch stem ends to promote bushiness and flowering.

ZELKOVA SERRATA Fast-growing, deciduous tree to 80′, shaped something like an elm. Grows in temperate climates.

Plant in early spring or fall in a sunny location in average soil. Water in dry spells. Fertilize every 3–5 years.

ZEPHYR LILY *Zephyranthes*, fairy lily. Bulbous plants to 15″ with graceful, lily-like, white or pink flowers in spring, summer or fall. Grows almost everywhere; also indoors.

Outdoors. Plant bulbs in spring 1″–2″ deep and about 6″ apart in well-drained, average soil with a little extra humus. Grow in sun or partial shade. Water in dry weather. Fertilize lightly when growth appears. Dig up in the fall and store in sand or peatmoss in a cool, dry place. In warm climates, the bulbs may be left in the ground.

Indoors. Plant in general-purpose potting soil. The bulbs should be set just below the soil surface and about 1″ apart. Place in a south window and water regularly. After blooming, gradually withhold water and then store dry in the pot for about two months. The bulbs can then be brought into bloom soon again.

ZINGIBER Ginger. Reed-like perennials to 3′, some with very showy clusters of bracts and flowers in whites and reds. Grows outdoors in warmest climates.

Grow like *alpinia* (which see). *Zingiber* dies to the ground in winter but comes back in spring.

ZINNIA Favorite annual with flowers in many colors. Plants grow to 3′, but some of the newest varieties are only a few inches tall. Grows anywhere.

For earliest bloom, sow seeds indoors about six weeks before last frost, or for later bloom, sow where plants are to grow outdoors. Grow in full sun in average, well-drained soil to which some humus has been added. Pinch stem ends of young plants to promote bushy growth. Keep watered. Fertilize when young plants are established and once more if you wish. Dust with sulfur (except in very hot weather) to control mildew. If possible, do not plant zinnias in the same place year after year.

ZUCCHINI *Cucurbita.* A thin, straight, dark-green summer squash. Grows almost everywhere.

Grow like squash (which see).

ZYGOPETALUM Epiphytic orchids to about 18″ with small sprays of winter flowers that are mixtures of green, brown and purple. A house plant.

For how to grow, see orchid.

BASIC GARDENING TASKS

HOW TO IMPROVE SOIL

The soil in the United States is extremely variable. Little of it is perfect. None of it is suitable to the successful growing of all types of plants. Therefore, steps usually have to be taken to improve or change it in some way.

What is "average soil"? "Average soil" is a term that appears throughout this book. In actual fact, because of the variations in our soil, there is no average soil. I use the term, however, to describe ordinary, run-of-the-mill soil that is neither extraordinarily fertile nor extraordinarily bad. Within wide limits, such a soil contains a certain amount of loam, humus and sand (the principal ingredients of garden soil) and it is more or less neutral (neither acid nor alkaline).

What is "general-purpose potting soil"? As described in the section titled How to Grow House Plants, general-purpose potting soil is a mixture of two parts loam, one part humus and one part sand.

How to lighten soil. "Heavy" soils feel heavy. They are dense and fine grained, tend to stick together, often contain considerable clay. Plants generally do not grow well in them. Therefore, they need to be lightened by mixing in sand and humus. How much depends on the original consistency of the soil and how light you want it to be.

How to give soil body. Sandy, gravelly soils are not favored by many plants because they do not hold moisture and perhaps do not supply adequate nourishment. You can improve matters by mixing in loam and humus.

How to improve soil drainage. The vast majority of plants die if their roots stand continuously in water. Consequently good drainage is a necessity in the garden. If your soil does not drain freely, dig extra deep holes for your plants and put 6″ or more of coarse gravel or hard cinders in the bottom. In boggy areas, in order to have good drainage, it is usually necessary to install drain pipes or ditches to carry off the water.

How to aerate soil. For plants, especially small plants, to grow well, their roots must receive oxygen. This means that the soil in which they grow cannot be packed as hard as a rock. One way to aerate all soils is to mix into them a large amount of humus. In addition, the soil should be broken up (cultivated) occasionally with a spading fork. In lawn areas, aeration is best done with a power machine that drives thousands of small holes into the ground.

How to change the pH of soil. The pH figure for a soil indicates whether it is acid, neutral or alkaline (sweet). A soil with a pH of 7.0 is neutral. A higher figure indicates an alkaline condition; a lower figure an acid condition. Most plants prefer a pH of 6.0–7.0.

You can measure the pH of your garden soil (as well as its fertility) with a soil-testing kit available from a garden supply store. However, it is better to send a soil sample to your state agricultural experiment station for a test.

To sweeten soil (raise its pH), apply hydrated lime or limestone (But never apply hydrated lime and fertilizer at the same time. Allow a week between applications.) Unless your soil is unusually acid, use about 1 lb. of ground limestone per 20 sq. ft., or 1 lb. of hydrated lime per 25 sq. ft. Repeat the application every 3–4 years.

To make soil acid (lower its pH), maintain a mulch of oak leaves or pine needles. To do the same job quickly, mix aluminum sulfate into the soil. Applying this to average soil at the rate of 1 lb. per 10 sq. ft. will lower the pH figure about one whole point.

How to fertilize. If you have access to a supply of well-rotted (not fresh) manure, use it. It may not be so potent in chemical content as dry fertilizers, but it does the job very well and in addition it contains a great deal of humus (few soils have enough of this). Dried manure that comes in bags is a good substitute.

If soil is known to be deficient in nitrogen, phosphorus or potassium—three of the chemicals essential to plant growth—you can buy the chemical in dried, powder form. Nitrate of soda is a commonly used source of nitrogen. It is very potent and fast-acting; however, it is somewhat difficult to handle and its effect is fleeting. Consequently, home gardeners generally use bone meal instead. This is not so rich in nitrogen, but it also contains phosphorus. It dissolves slowly and is therefore nourishing to plants over a long period of time.

Superphosphate is the outstanding source of phosphorus if phosphorus alone is needed in a soil. Muriate of potash supplies potassium. (Wood ashes also supply small amounts of potash.)

Instead of using three separate chemicals on their plants, however, most gardeners find it much more convenient to use a balanced commercial fertilizer which supplies all three chemicals at one time. Many different balanced fertilizer formulas are available. These are, by law, clearly printed on the bag in the following fashion: 5–10–5. In all these formulas the first figure indicates the percentage of nitrogen contained in the fertilizer; the second figure indicates the phosphoric acid content; and the last figure indicates the potash content.

Which fertilizer formula you should use depends on the natural chemical content of your soil (for example, if it is deficient in phosphorus, you need a fertilizer with a high phosphoric acid content) and on the needs of the plants being fertilized (grass and plants grown for their foliage need more nitrogen than anything else while root vegetables usually want potash). As a practical matter, however, an easier and usually adequate procedure is to buy what I call in this book a "general-purpose fertilizer". This is any fertilizer that comes in a bag labeled "for lawns and gardens" or something to this effect. Typical general-purpose fertilizer formulas are 5–10–5, 8–16–8, 5–8–7, 4–8–7.

One other point to note about balanced commercial fertilizers is that some use a nitrogenous material that releases nitrogen quickly to the soil while others use a material that releases nitrogen slowly, over a longer period of time. The first is called a fast-acting fertilizer; the second, a slow-acting, or slow-release, fertilizer.

Slow-acting balanced fertilizers are commonly used only on grass. Fast-acting fertilizers are of the general-purpose type and are used on grass and everything else.

When using a fast-acting general-purpose fertilizer, remember that it can burn foliage and small roots. If you get any on leaves, wash it off thoroughly. When applying it to lawns, water it in immediately and well. In the garden, mix it thoroughly with the soil (or scratch it well into the surface) and then water heavily.

To feed trees (except those in orchards, where the fertilizer is just scratched into the soil), a special technique is used. With a crowbar, make a ring of holes in the ground around the tree directly under the branch ends. The holes should be about 18″ deep and 18″–24″ apart. Half-fill each of the holes with ordinary general-purpose fertilizer (or a general-purpose mixture that has a high nitrogen content; for example, 10–6–4 or 10–8–8); then fill in the rest of the way with soil, and water deeply. Use 3 lb. of fertilizer per inch of trunk diameter.

How to improve soil with "green manure." One of the best and cheapest ways to improve soil (especially in suburban areas where builders have stripped off much of the topsoil) is to grow on it a succession of cover crops which are plowed under. The principal purpose of these is to add humus to the soil. Some plants also add nitrogen.

Plants used for green manuring include buckwheat, rye grass, clover, vetch, alfalfa, lespedeza, cow peas and soybeans. All but the first two are legumes which absorb nitrogen from the air and put it into the soil.

One way to handle green manure is to sow soybeans in the spring at the rate of 3 lb. per 1000 sq. ft. Plow these under in early September and then sow annual rye grass at the rate of 5 lb. per 1000 sq. ft. Plow this under the next spring and plant a new crop of soybeans.

Another way to handle green manure is to sow annual rye grass in September and feed it with a general-purpose fertilizer. This produces a strong, thick grass cover before winter. The next spring, apply calcium cyanamide to the soil at the rate of 10 lb. per 1000 sq. ft.; then plow under the rye grass. The cyanamide hastens decomposition of the grass and puts nitrogen into the soil.

Then plant a second crop of rye grass in the same way (but you can omit the fertilizer) and plow it under when it is thick. Then plant a third crop and plow it under in August or September. (This time you can omit both the fertilizer and the cyanamide.)

HOW TO COPE WITH NATURE

Light, moisture, frost, winter and wind all have an important effect on the growth and growing of plants. Here are a number of rules in connection with them that you should remember.

Light. Try to give plants the full amount of sunlight they require to grow and perform properly. If they need a full day of sun, plant them where they will get a full day of sun. If they need shade, give them shade. Trying to make them grow with more or less sunlight than they actually need is hopeless.

Don't grow plants too close together. If you do, they will shade one another and eventually some of them will die for lack of light (and moisture) while others will become tall and gawky as a result of their efforts to reach up to the sun.

Locate trees, large shrubs and vines so that, even when they grow large, they will not cast your entire yard in shadow and make it impossible to grow some of the smaller plants that need sunlight.

Water. When watering with a hose or watering can, don't just sprinkle the surface of the soil. Apply enough water so that it penetrates deeply into the soil. A few deep waterings are much more beneficial to plants than numerous shallow waterings.

Deep-rooted plants do not have to be watered so often as shallow-rooted plants; but when they are watered, they must be watered very thoroughly to make sure that the moisture gets down to the roots.

In cold climates, water trees and shrubs—especially evergreens—heavily before the ground freezes in the fall. Mulching the soil helps to hold in moisture both winter and summer.

Water dormant plants sparingly, if at all.

If possible, don't grow small plants within reach of the roots of large trees. In the competition for the available moisture in the soil, the trees are bound to win out.

If plants are growing outdoors in pots, tubs, window boxes, etc., check the water supply every day. In hot weather especially the soil dries out rapidly—often in only a few hours.

Water plants growing outdoors early in the day rather than in the evening.

Remember that plants growing in sun require more water than those in shade; that the more humusy a soil is, the longer it holds moisture; that sandy and gravelly soils need more frequent waterings than humusy soils (they also need more fertilizer, because heavy watering washes the nutrients out of the soil).

Frost. Protect small plants from late spring frosts and early fall frosts by covering with Hotkaps, baskets, tents of newspaper, etc. Hanging heat lamps several feet above cherished plants in a small garden also gives good protection. In orchards, protection is achieved with smudge pots and by constant movement of the air with huge fans.

Don't locate plants which are very susceptible to frost in low spots, where the frost settles first. Grow on high ground.

Winter. Mulch the soil around small plants after it has frozen to prevent them from being damaged by alternate freezing and thawing of the soil.

Protect fall-planted evergreens by spraying the foliage with a plastic which retards the loss of moisture through the leaves. An alternative is to erect a screen of burlap on the windward side of the plants.

To prevent heavy snows from breaking limbs of trees with weak wood, support large limbs on sturdy timbers.

To prevent foundation shrubbery from being crushed by snow sliding off steep roofs, install metal snow guards on the roof. In addition it may

be advisable to build a cover of laths or burlap over dense evergreens, such as box.

See How to Grow section for other ways to protect specific plants in winter.

Wind. In areas with strong prevailing winds, avoid locating plants directly in the path of the wind. Small ones are battered to bits; even large trees become badly distorted.

HOW TO PROPAGATE PLANTS

There are a number of ways to reproduce plants. Not all are easy; but if you want to save money or if you want to raise a particular plant that you see in a friend's yard, it is worthwhile taking a chance even on the hardest.

Seeds. See How to Grow Flowers and Vegetables from Seeds.

Stem cuttings. A stem cutting is the end of a stem growing on a vigorous plant. Many plants may be propagated by such cuttings.

Make stem cuttings at any time when the plant to be propagated is growing strongly. As a general rule, June, July and August are best (but for some subtropical plants, the schedule may be quite different). In most cases, the stem to be cut should be fairly young (a greenwood cutting); but in some cases, it should have matured a little (the stem is said to be semi-ripe or half-ripened). In no case should the stem be flowering.

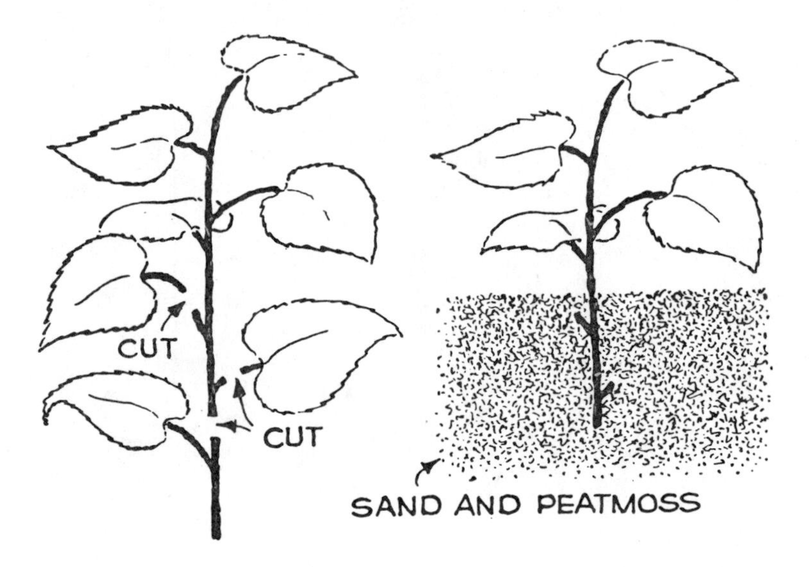

The cutting should have about four leaves. Make a slight diagonal cut ½″ below a leaf node, or joint, with a very sharp knife. Then remove one or two of the bottom leaves. If the cutting cannot be planted immediately, wrap it carefully in damp cloth or newspaper.

An easy way to root cuttings from a number of plants is to insert the bare ends of the stems in a jar of water. For fastest action, burn a 60-watt light 18″ above the cuttings for about 18 hours a day. Some of the plants that make a good response to this treatment are *begonia*, geranium, *coleus*, *kalanchoe*, *peperomia*, *philodendron*, *vinca* and *gardenia*.

The other way to root stem cuttings is to plant the bare parts of the stems in a pot or box filled with a half and half mixture of damp sand and peatmoss. First dip the base of the cuttings in water and then in a hormone powder which encourages root formation. (Ask the garden supply store which strength powder to use for each cutting.) Then insert the stems in the "soil," firm well and water. Keep the cuttings in a good light, but not sunlight, and in a moderate—about 70°—temperature.

To prevent the cuttings from wilting, keep the soil damp (but not soggy) at all times and spray the cuttings with a fine mist of water every day. The alternative—which is recommended—is to plant the cuttings in a flat. Bend stiff wire mesh over the flat like a quonset hut; it must not touch the cuttings. Then wrap a sheet of polyethylene film tightly over the top and

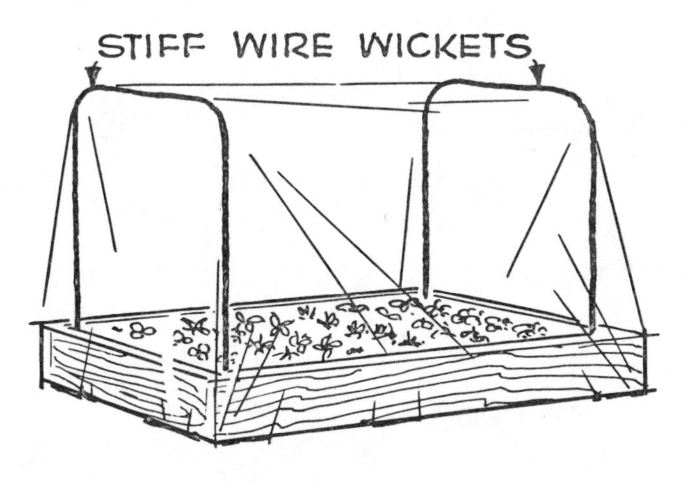

around the ends and sides of the flat, and thumbtack this to the bottom edges of the flat. This will hold in water vapor and thus make it unnecessary to worry about watering or spraying the cuttings. However, keep an eye out for fungus on the cuttings or on the soil in the flat. If you see any sign of this, you must open the polyethylene tent about once a week and spray the cuttings with a fungicide such as captan.

When the cuttings have rooted, transplant them into real soil and keep in light shade for about a week. If you rooted them under polyethylene, it is also advisable to keep them covered with the film. But remove the film for a longer and longer time each day until the cuttings no longer threaten to wilt.

Leaf cuttings. Some house plants are easily reproduced from large, young leaves. This is, for instance, a favorite way of starting new African violets. All you do is remove from a vigorous plant a leaf with a stem about 2″ long. Fill a jar with water and place over the top a piece of wax

paper with a slit in the center. Insert the violet stem through the slit into the water.

Another way to handle African violets as well as gloxinias, *peperomia,* some *begonias, sedums* and *echeveria* is to cut off a leaf with just a bit of stalk and insert the stalk in moist sand or vermiculite. The rooting medium must not be allowed to dry out.

Other plants, such as snake plant, are propagated in slightly different ways from leaves. These are described in the How to Grow section at the appropriate plant entry. In all cases, cover the leaves lightly with polyethylene to prevent them from drying out.

Layering. This is an easy way to propagate vines, trailing and creeping plants, plants putting out runners and even a number of upright plants. Generally, with thin-stemmed plants, such as ivy and periwinkle, all you have to do is pin a portion of a stem (but usually not the tip) down to

cultivated soil with a wire or stone; cover with humusy soil, and keep moist. When roots form, sever the newly rooted section from the parent, and plant it.

Larger plants, such as various forms of rhododendron, magnolia and viburnum, are layered by making a slight diagonal cut in a branch about

18″–24″ from the tip. Insert a toothpick to hold the cut open. Then bend the branch to the ground, weight with a stone and cover with about 3″ of soil.

Layering is best done in the spring. With many plants, roots form by

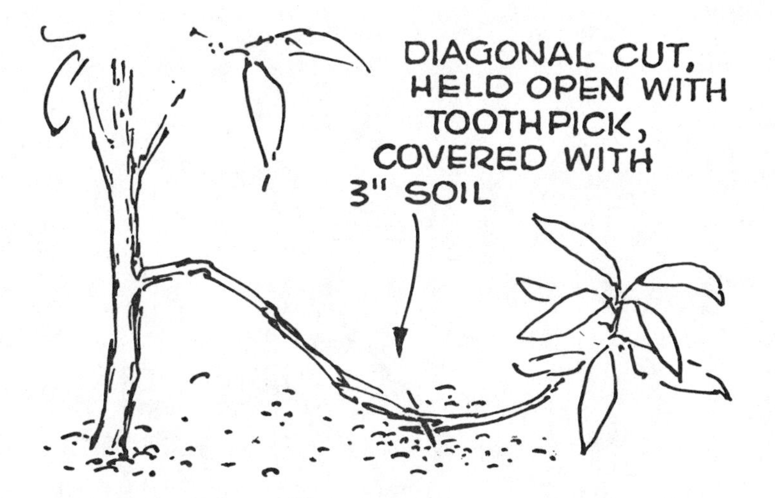

autumn. Some, however, take longer to develop a root system sturdy enough to permit moving, for several years.

Air-layering. This method is used to propagate many plants, such as dogwood, lilac, holly, *oleander*, *gardenia*, rubber plant, *dracaena*, etc. Use the simple air-layering kits that are available at garden supply stores.

Air-layering is best done before the buds open. Make a shallow notch in a year-old stem, or slit the stem up and down for about 1″ (without cutting through it), and remove the bark from the flap. Dust the wound with a hormone powder that comes with the propagating kit. Then saturate a wad of sphagnum moss with water and squeeze it as dry as possible. Wrap this around the wound and over-wrap tightly with plastic film. Seal the joint in the film and also the ends with cellulose electrical tape so that moisture cannot escape. Roots will form at the wound, after which cut off the stem below the wound and plant it.

Suckers. Plants which throw up suckers from the ground, such as lilacs and locusts, are easily propagated simply by cutting out the suckers with roots attached and planting elsewhere.

Bulbs and corms. Almost anyone who has done any gardening at all knows that when daffodil bulbs that have been in the ground for a couple of years are dug up, a host of little bulbs are found alongside. Similarly, when *gladiolus* corms are dug up in the fall, tiny cormels are found attached. If these little bulbs and corms are planted, they will in a few years' time develop into large, flower-producing bulbs and corms.

Of course, not all bulbous and cormous plants are as obliging as daffodils and *gladiolus*, but those that are are readily propagated by their offsets.

Rhizomes and tubers. Plants growing from these are commonly reproduced by cutting the rhizomes and tubers into smaller pieces, each of

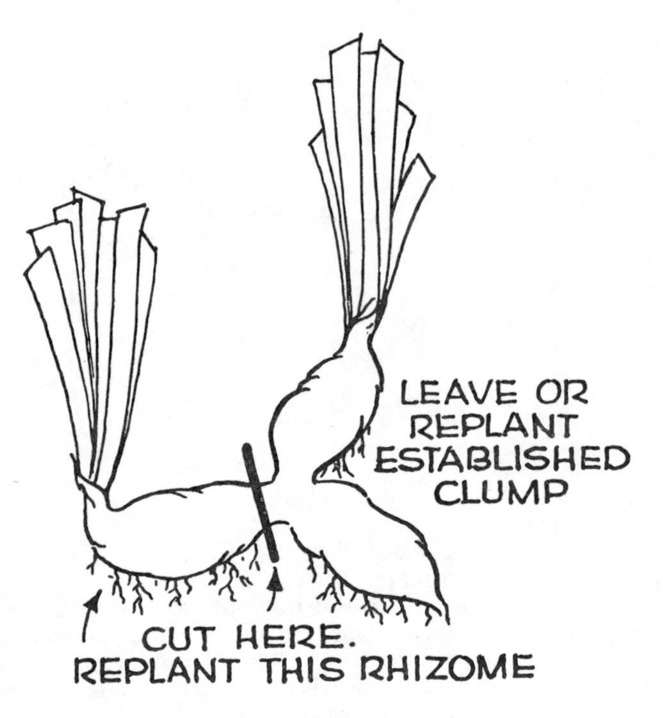

which contains one or more buds, eyes or growing points. Dust the cut surfaces with powdered sulfur to prevent rot.

Division. The surest way to develop new plants of perennial flowers and ferns is to dig up a large clump and divide the roots into smaller sections, each having one or more stems. Some plants are easily pulled apart by hand. Others need to be cut apart with a knife or sharp spade, or even pried apart with a couple of spading forks placed back to back.

As a rule, early-blooming plants are divided in the fall; late bloomers in the spring. But there is nothing sacred about this timing.

HOW TO GROW FLOWERS AND VEGETABLES FROM SEEDS

Note. Flowers and vegetables are the plants most commonly grown from seeds by non-professionals; therefore, this chapter is limited to them. However, in this book, you will occasionally find the suggestion that other types of plants may be grown from seeds. In those cases, if special seeding methods are called for, they are described at the entry. If special methods are not called for, the instructions which follow apply.

Outdoors. After thoroughly spading the soil and adding whatever enrichment is necessary, pulverize it thoroughly and rake smooth. If the soil is very dry and the seeds to be planted are very small, water the soil well. The reason for this is that if you watered after the seeds were sown, they would probably be scattered far and wide.

Because of the danger that the seeds and tiny sprouting plants may be killed by a fungous disease known as damping-off, treat the seeds before sowing with a special powder and shake well. An alternative is to water the seeds after they have been sown with a chemical called Pan-O-Drench.

Seeds are planted either in rows (which may be called drills or furrows) or they are scattered over a wide area for an informal effect. In either case, the general rule is to cover the seeds with soil equal to the diameter of the seeds. However, some vegetable seeds may be planted much deeper than this; and minute seeds, like those of petunias, are not covered at all (they are simply pressed lightly into the soil surface).

When the seeds are sown, tamp the soil over them lightly with a rake or flat piece of wood. Then, if the soil is not already decently moist, water with a fine spray.

Thinning out of the seedlings that come up should start as soon as the plants are 1″–3″ tall.

Seeds sown in flats, coldframes or hotbeds are handled in the same way as those sown directly in the garden.

Indoors. Seeds are started indoors for house plants and also in order to get a headstart on the summer. For instance, in the New York City area, if you wait until the danger of frost is past before sowing marigold seeds outdoors, you won't have flowers until about July 15. However, if you start the seeds indoors 4–6 weeks before the last frost, you should have flowers outdoors about June 15.

Indoors, seeds should be sown only in soil that has been sterilized (see How to Grow Plants Indoors) or in some inert material such as clean sand, vermiculite or sphagnum moss. This precaution will prevent damping-off. If soil is used, the seedlings can continue to grow in it until they are ready to be transplanted outdoors. If ordinary sand, vermiculite or sphagnum moss is used, the seedlings will receive no nourishment so it will be necessary to transplant them to pots or flats of sterilized oil soon after they have formed roots. However, there are on the market special seed-starting kits which make the transplanting of the tiny seedlings unnecessary. These consist of small plastic trays which are filled with vermiculite mixed with plant food. Seedlings get enough nourishment from this mixture to continue in the trays until they are ready to go into the garden outdoors.

Seeds can be started indoors in any container that is 2″ or more deep. Plastic trays like those mentioned above are very good, but you can also use wood boxes, tin cans, flower pots, etc. In all cases, however, the containers must be scrubbed with soap and water before they are used, and they should have holes in the bottom for good drainage.

Sow the seeds in rows or scatter them over the surface of an entire container. Then place the container in a bowl of water until the soil surface is damp. Cover lightly with aluminum foil or wax paper to hold in moisture and keep in a bright (but not sunny) place until the seeds sprout. Then move into a sunny window and turn the containers daily to prevent the little plants from growing lopsidedly toward the sun. Thin the plants when they are about 1″ tall and water regularly (but don't let the soil get soggy) until it is time to move them outdoors.

HOW TO TRANSPLANT

Transplanting is to plants what surgery is to humans: It may improve them, but at the time of transplanting and for some days afterwards it puts them in a mild state of shock. They may recover and they may not.

It follows that transplanting must be done with care and that the patient should be in good health before it is uprooted and moved. (However, sick plants often have to be transplanted.)

Small seedlings. Transplant at any time necessary.

If the plants in an entire container (flat, pot, pan, etc.) are to be transplanted, loosen the soil in a chunk from the container, and drop the soil

chunk bottom-side down on a table. This breaks up the soil and makes it easy to lift out the individual seedlings.

If only some of the plants in a container are to be transplanted, moisten the soil slightly so that it is easier to handle; then gently dig up the seedlings with a blunt knife.

Poke a hole in the soil in the new pot or flat with a knife, pencil or your finger. Hold the seedling at the proper depth in the hole (it should not be planted any deeper than it formerly grew) and press the soil around it. Water with a starter solution such as Transplantone. Don't expose the seedlings to sun for a day or two.

Large seedlings. Transplant these outdoors when weather is reliable and frost danger is past. Transplant, if possible, only on a cloudy day or in later afternoon; otherwise the plants may wilt badly in the sun.

Water the plants several hours before they are to be moved.

Turn over the soil where the seedlings are to be planted and improve it as necessary (but not with fertilizer). Scoop out ample holes for the plants with a trowel.

Dig up the plants with as much root as possible. Don't let them stay long out of the ground. Set them into the newly dug holes and spread out the roots. Then fill in around them and firm the soil. As a rule, the plants should be set at the same depth as they previously grew. Form a depression in the soil around them and fill this with water. When this has soaked

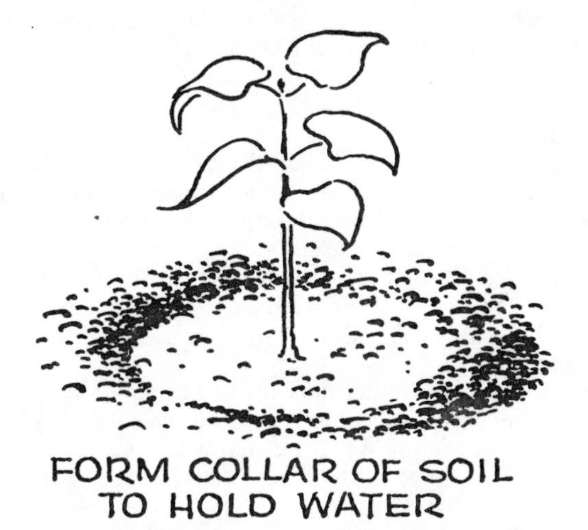

FORM COLLAR OF SOIL TO HOLD WATER

in, it's a good idea, though not essential, to pour in about one-half cupful of a starter solution (Transplantone, for example). Keep the plants watered in the days that follow until they no longer show any signs of wilting.

Don't give newly transplanted seedlings any general-purpose fertilizer for about a week. Then mix a small amount into the soil around them and water it in well. Don't let any fertilizer stand on the leaves.

Seedlings that were previously grown in pots, plant bands, paper cups or tin cans are more safely transplanted than those grown in a flat or seedbed because they have a ball of earth around the roots. Just tap the sides of the container to loosen the rootball, and set the rootball in the planting hole level with the surrounding soil. Don't break the rootball. Water well. A starter solution is unnecessary.

Plant divisions and established perennials, biennials, ferns, etc. Plants that have been growing outdoors can be transplanted in the spring or, in many cases, in early fall. Spring planting can be done before the last frost (it is necessary to protect the plants with newspapers or Hotkaps only if you know a very severe frost is coming). Fall planting is often preferred in warmest climates, but is never advisable in the coldest.

Move the plants in either case at any hour of the day. Since they are already well developed, they are not bothered by sun.

Plants that have been growing indoors should not be transplanted outdoors until all danger of frost is past. If they are to grow in the sun, they should be gradually acclimated to it. That is, start out by screening them from the sun for almost the entire day; then gradually reduce the time of screening. (Even though plants have been growing in a very sunny south window, they are not accustomed to the full blast of the sun outdoors.)

Handle all large plants or plant sections as you do large seedlings (see above). Application of a starter solution is not necessary but is helpful. Hold off general-purpose fertilizer for 3–4 days.

Plants growing from bulbs, tubers, corms, rhizomes. Specific instructions are given for individual plants in the How to Grow section. But note the following general rules:

Unless you are naturalizing quantities of bulbs in fields and woods, try to spade up and pulverize the soil in which the bulbs are to be planted.

Mix a small amount of general-purpose fertilizer or, better, bone meal into the soil below the bulbs, tubers, etc.

Make sure that the bulb, tuber, etc. rests squarely on the bottom of the planting hole. In other words, eliminate air spaces under it. To make sure you do this, dig holes with a trowel or spade. Don't use a dibble (a round, pointed stick).

Trees and shrubs. Plant deciduous plants early in the spring just before they leaf out, or in the fall after they drop their leaves. Plant evergreens a little later in the spring but before growth has started or in early fall (about September 1). Note, however, that in very cold climates spring planting is almost always preferred.

Trees and shrubs bought from a local nursery or plant store usually come with a large rootball wrapped in burlap. These plants are said to be balled and burlapped, or they are labeled "B & B."

The hole in which you plant a tree or shrub should be about 1′ wider than the rootball and 6″ deeper (unless the subsoil is hard clay, in which case it is necessary to go much deeper to assure good drainage). As you dig the hole, place the topsoil in one pile, the subsoil of the hole in another. Mix the topsoil with a couple of handfuls of general-purpose fertilizer and with humus and/or sand (this is not always necessary). Then toss it back into the hole. Set the rootball on top of this layer of soil, making sure that when the hole is filled completely the plant will be at the depth that it grew in the nursery. Cut or untie the burlap from around the stem of the plant but don't bother to remove it because it will soon rot. Then fill in around the rootball with the remaining soil. The topsoil should go in first; the subsoil, last. As you fill in, tamp the soil firmly with your foot. When the hole is finally filled, build a 3″–4″ high collar of soil around the plant. It should be at least as wide as the hole you dug. Fill the depression within the collar with water and refill it every three days for two weeks. Thereafter the plant should need water only in dry weather or according to its own peculiar requirements. Note, however, that all trees and shrubs—especially evergreens—should be watered deeply before the ground freezes if you live in a cold climate. Note also that fall-planted

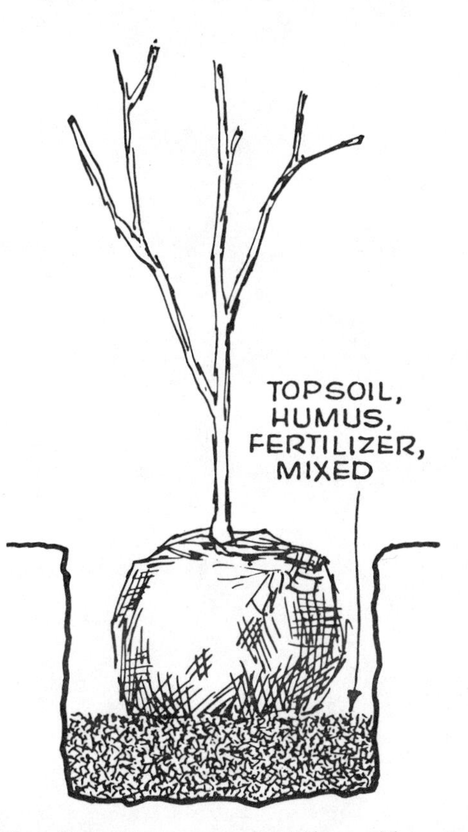

evergreens are benefited if you spray the foliage with water frequently to prevent loss of moisture by transpiration. (An even better way to treat them is to spray the foliage with a special liquid plastic available at garden supply stores.)

Additional fertilizer should not be needed for a year.

Trees should be staked to hold them upright. A 1″ stake will support trees to about 6′ in height. Larger specimens or those with heavy tops should be supported with three guy wires looped around the trunk and fastened to stakes in the ground. The wires should be covered with sections of garden hose where they pass around the trunk.

Wrap the trunks of fall-planted trees with burlap to protect them against sun-scald. The covering, usually applied in spirally wound strips, should extend from the ground up to the bottom branches. Keep it on for a year.

Trees and shrubs purchased from nurseries that sell by mail usually

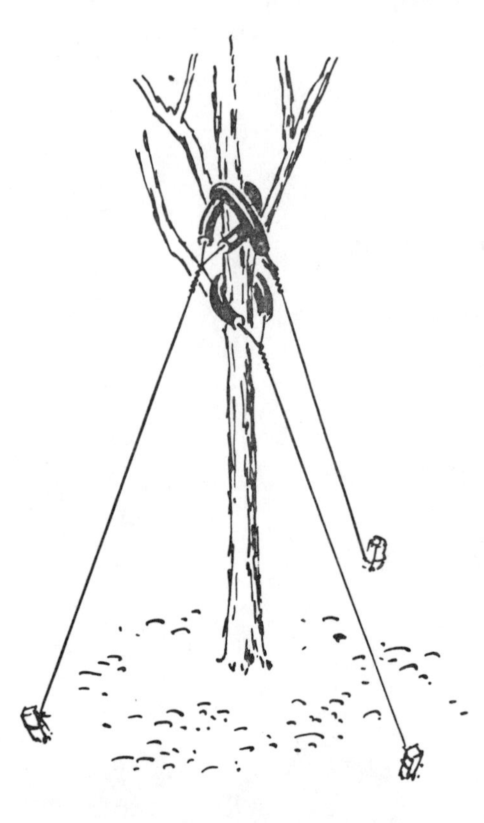

come with bare roots. These should be planted as soon as they arrive; if this is impossible, don't open their wrappings until you are ready to plant.

The technique of preparing the planting hole is the same as for B & B plants. Clip back broken roots and spread the roots out naturally. Don't cramp them into a tight hole if you want the plant to amount to anything. Then fill in with soil, water, stake and wrap with burlap.

One additional step called for is to prune the branches slightly to compensate for the pruning of the roots.

Trees and shrubs that you dig up and move yourself require even more attention.

Ideally, if you know that you are going to move a woody plant some day, you should cut the roots in a circle around it a year before moving day. Do this with a very sharp spade held vertically and plunged to its depth in the ground.

Usually, however, most people move shrubs and trees pretty much on the spur of the moment. Therefore, no preparatory root-pruning is possible. All you can do is soak the soil thoroughly a day before the plant is to be dug up. Then dig a trench in a circle around the plant and cut horizontally under the roots. Then slip burlap under the rootball, wrap it tight and lift out the tree.

But be warned that this is difficult, back-breaking work. It is essential that you avoid cutting the roots as much as possible. It is hard to hold a ball of earth together and to wrap it. And the labor of lifting and moving the tree is considerable if the plant is of any size. In other words, you almost certainly need strong assistance. It is really better to let an expert do the job for you.

Let's assume, however, that you do the job yourself. In that case, handle the plant like a B & B specimen (above). Just be sure that you clip off cut and broken roots. Then cut back the branches about one-third to compensate for the loss of roots.

Vines. Vines are usually vigorous growers that can be cut back hard without great damage. If possible, transplant them with a ball of earth; but if not, don't worry too much. Just take as much root as you can. Handle as you would a shrub or tree and on the same schedule.

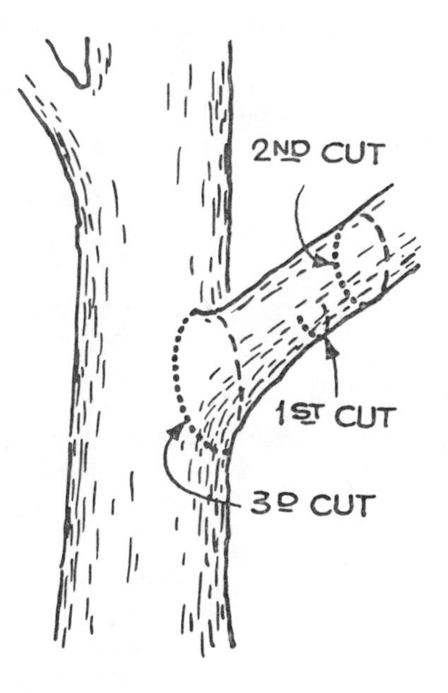

HOW TO PRUNE PLANTS

The purposes of pruning are to eliminate dead, diseased and damaged wood; to keep a plant at manageable or a specially desired size; or to improve flowering or fruit production.

Throughout the How to Grow section you will find instructions—often very detailed—for pruning specific plants. Here are some general rules that should also be observed:

Tools and technique. Always use sharp tools. The more you mangle a plant, the more chance that disease will enter.

Dress all pruning wounds over $1\frac{1}{2}''$ in diameter with tree paint. Dress smaller wounds if the plant you are pruning is sickly.

Disinfect tools used on diseased wood before using them again.

When removing a tree limb, first make a cut part way through the underside of the limb several inches out from the trunk. Then saw through the limb from the top at a point a little further out from the trunk. Then saw off the stub as close as possible to the trunk. The purpose of all this is to avoid tearing the bark from the trunk below the limb and possibly even splitting the wood.

To prune shrubs that send up a lot of stems (for example, forsythia or climbing roses), cut the old and weak stems back to the ground occasionally. This produces a far more effective plant than if you simply chop back the ends of the branches.

To make a plant bushier, nip off the growing tips of stems and branches.

To make a plant grow outward instead of upward, cut off the uppermost growing point, or leader. Note, however, that conifers will develop a new leader.

To make a plant grow upward instead of outward, cut out side branches near the bottom.

Timing. Prune non-flowering deciduous plants at any time. Late winter, however, is probably the most convenient time.

Prune plants that flower in winter and spring right after flowering.

Prune plants that flower in summer and fall in late winter or early spring.

Prune plants that flower off and on throughout the year after one of their flowering spurts during warm weather.

Conifers. As a rule, remove only short lengths from the ends of branches. Otherwise, you wreck the symmetry of the plant.

Prune in late spring or early summer—never in winter.

Don't cut a branch back beyond the last needle if you want the branch to live.

HOW TO GROW HOUSE PLANTS

Several hundreds of plants ranging from annuals to small trees are easily grown in the house. Each has its own special requirements, but for all of them the basic method of treatment is essentially the same.

Containers. The best containers are ordinary red clay flower pots with drainage holes in the bottom. They are inexpensive and attractive when clean, and because the clay is porous, the soil is aerated to some extent and does not become waterlogged.

Plastic pots, also with bottom drainage holes, are almost as good. Their principal advantages are their cleanliness and light weight. But they do not allow the soil to breathe in the way that clay pots do.

Many other types of containers are used for house plants, but if they do not have drainage holes, they should be used only as a last resort because there is danger that the soil will become waterlogged and sour. No plant does well under those conditions; in fact, the odds are that it will die unless you soon come to the rescue.

Soil. Most house plants do well in soil composed of two parts loam, one part humus (peatmoss, leafmold, etc.) and one part coarse sand. Throughout this book I refer to this mixture as "general-purpose potting soil."

Sterilize the soil before planting in it. This is not an absolute essential, to be sure; but it is a wise precaution that may make the difference between a healthy plant and a dead one. To sterilize soil, simply moisten it well and bake in an oven at 200° for two hours. Do not plant in the soil for a couple of days thereafter.

If you wish to avoid the nuisance of mixing and sterilizing soil, buy ready-mixed potting soil at a garden supply store or the five and ten.

Fertilizing. All house plants, even the most durable and undemanding, need an occasional dose of plant food. Use any balanced commercial fertilizer you like. However, those made especially for house plants are preferred; and of these, the liquids are probably the easiest to use. Follow the directions on the label carefully. Never make a stronger solution—or use more of any kind of fertilizer—than the manufacturers recommend.

Apply fertilizer when plants are starting into growth and making growth, not when they are resting. The frequency of application depends on the type of plant. If using liquid plant food, apply it *after* you water the plant.

Watering. All house plants—even desert cacti—need water while making growth; but when they are resting, the amount must be reduced. The frequency of watering during the growing season varies with the type of plant. Many should be watered only when the surface of the soil feels dry. Many others should be kept constantly moist, but almost never soggy.

There are two ways of watering plants grown in containers with bottom drainage. One is to pour the water on the soil. The other is to set the pot in a bowl of water up to half its depth. Top-watering is easier and neater, but you must make sure you apply enough. Keep adding water until it runs out of the bottom of the pot. If you water plants from the bottom, let them stand in the water until the surface soil becomes moist.

While it is important that the soil for growing plants be wet through when water is applied, you almost never should allow potted plants to stand in water once the soil is wet. This causes the soil to become waterlogged.

Light. Some house plants need all the sun they can get. Others need sun only half the day or even less. Still others don't need any sun at all.

The sun lovers should be grown in a south window. Those that want partial sun are grown in east or west windows. Those that don't need sun at all are grown in north windows.

If because of the orientation or design of your house or because of trees and buildings around it, you cannot give plants as much light as they need, you can compensate for this lack by growing them under fluorescent light. Ready-made fixtures are available, or you can have two 40-watt daylight fluorescent tubes mounted about 1′ apart on a board. Hang the lights above the plants on adjustable chains.

As a very general rule, the lights should hang about 1′ above the tops of the plants. Give the plants a total of 12–14 hours of light a day. This can be electric light or a combination of electric light and natural light.

Temperature. For most house plants, 68°–70° is the maximum desirable daytime temperature. At night, readings down to 60°–65° are desirable.

Many plants, however, need lower temperatures to thrive. Only a hand-

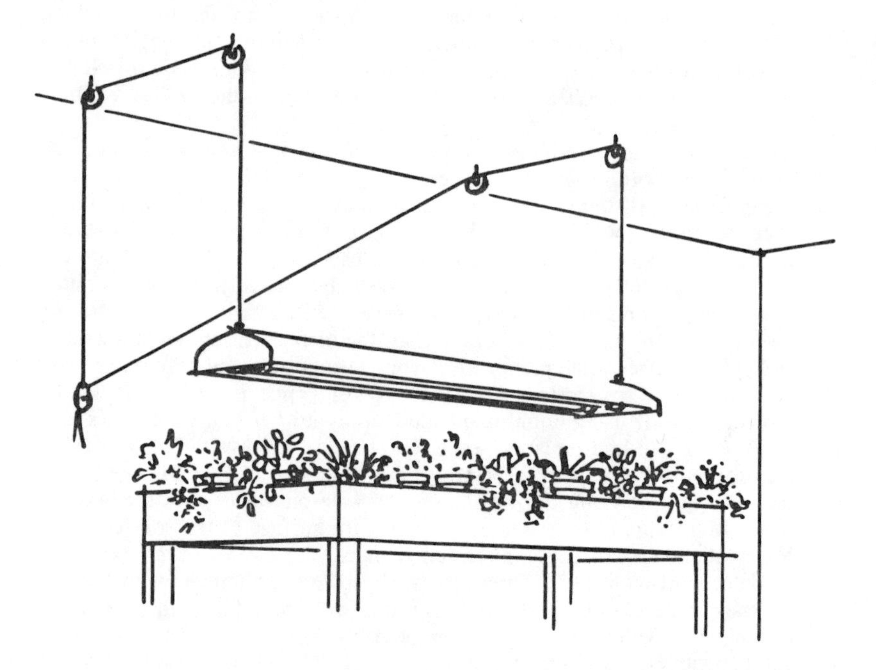

ful prefer higher temperatures. These special cases are noted in Section 1 of the book.

In cold climates, plants growing close to windows may be damaged by the cold seeping through the glass. To protect them, move the pots further away from the glass or put layers of newspapers or cardboard between plants and glass on coldest days.

Humidity. The average house has a dry atmosphere that is discouraging to a great many plants. However, their need for humidity can be satisfied to some extent.

The simplest solution—though it is not so effective as some people think—is to place a deep layer of pebbles in a saucer, pour in water just to the top of these, and then set the pots on top. The water level in the saucer, must, of course, be maintained.

A better way to give plants humidity is to syringe the foliage with water. Use a fine spray to avoid battering the flowers and leaves. After spraying, do not set the plants in a sunny window until the water on the leaves has evaporated; otherwise, the foliage may be burned.

Ventilation. Like humans, plants do not enjoy a stuffy atmosphere. Even in coldest weather you should occasionally ventilate the room in which they are kept. Do this in such a way that the plants are not exposed directly to drafts.

Rest periods. All plants have periods when they are dormant to some extent. That is, they stop growing and blooming. When this happens, don't worry (unless, of course, it is not the normal period of dormancy). Just stop fertilizing the plants and reduce the amount of water you ordinarily give them.

Transplanting. When house plants begin to crowd the pots they are in, they usually stop growing well. Then all you have to do is remove them from the pots and replant them in the *next-size-larger* pot. When doing this, put as much new soil around the roots as you can. The best time to make the shift is just before the plants begin to make new growth.

If a plant is already in as large a container as you can accommodate in your house, the best treatment is to remove it from the pot, scrape some of the soil from the rootball, trim the roots with scissors and replant in the same pot with as much fresh soil as you can work in. An alternative to this is to divide the plant.

HOW TO FORCE BULBS FOR INDOOR BLOOM

Forcing is the gardener's word for making plants bloom ahead of schedule. The most common and effortless forcing is done with *forsythia* and pussy willow branches, which are cut in late winter and brought indoors to bloom weeks before flowers appear outdoors. But an even more delightful (though somewhat more difficult) kind of forcing can be done with daffodils, tulips, hyacinths and various other less important spring bulbs. Here is how you go about it:

In October or November pot the bulbs in general-purpose potting soil. Daffodils and tulips are usually planted 5–8 bulbs to a large pot. Hyacinths are planted three bulbs to a 6″ pot or one to a small pot. The bulbs should be set so that their growing tips just show above the surface of the soil. When several bulbs are planted together, they may almost touch one another.

For a succession of bloom, plant several pots at the same time. The time of bloom is controlled, not by the time of planting, but by the time you start to force the bulbs.

Water the newly planted bulbs until the soil is damp through. Then place in a corner of your garage or a cold basement and cover with burlap. The bulbs should not be allowed to freeze, and the temperature should not rise above 40°. Don't let the soil dry out completely.

After the bulbs have been in "cold storage" for about six weeks, check to see how well their roots have grown. If the roots are coming out of the hole in the bottom of the pot, the bulbs are ready for the next step. If the roots have not appeared, turn the pot on its side and rap it with a stick of wood until the ball of earth can be slid out. If the ball is covered with tiny, white roots, you are ready to proceed with the forcing operation. But if the root growth is sparse, replace the rootball in the pot, put the pot back in its corner and check it again in another 10–14 days.

As soon as root growth is well developed in a pot, the pot can be moved into a dark closet at a temperature of about 50°–55°. Now the tops of the bulbs will really start growing. Keep the plants in the closet for about three weeks, or until the leaves and flower stalks are 6″–8″ tall. (They will be white in color, but don't worry about this.)

Now bring the pots out of the closet and place them in a warm, sunny window. The foliage will turn green and start to grow very fast; and within about three weeks the bulbs should start to bloom.

The whole forcing procedure takes about 12 weeks from the time the

WHEN EARTH BALL IS COVERED WITH ROOTS BULBS ARE READY FOR NEXT FORCING

bulbs are potted until they flower. But you can have a succession of bloom over a period of months by leaving some of the pots in "cold storage" and bringing them out at 10–14 day intervals. (In other words, as long as the potted bulbs are held in cold storage, they will not develop very much top growth.)

The following are a few of the varieties of bulbs which are especially suited to forcing:

Daffodils: Golden Harvest, Rembrandt, Music Hall, Spring Glory, Beersheba, Mrs. Ernest H. Krelage, Carlton, Scarlet Elegance, Flower Record, Mrs. R. O. Backhouse, Lady Kesteven, Edward Buxton, La Riante.

Hyacinths: Tuburgen Scarlet, La Victoire, Jan Bos, Distinction, Princess Margaret, Pink Pearl, Lady Derby, Anne Marie, Perle Brillante, Ostara, Myosotis, Bismarck, Orange Boven, Prince Henry, Arentine Arendsen, L'Innocence, Edelweiss.

Tulips: Single and double early types; also the Mendels.

GLOSSARY OF TERMS AS USED IN THIS BOOK

Air-layer. See How to Propagate Plants.
Average soil. See How to Improve Soil.
Bract. A colored, leaf-like, plant organ resembling a flower. The big red flowers of the poinsettia, for example, are actually bracts; the real flowers are the little things in the center of the bracts. Similarly, the white and pink flowers of dogwood are really bracts.
Broad-leaved evergreen. A plant with large leaves—not needles—which it holds throughout the year. See evergreen.
Bromeliad. A member of the Bromeliaceae family of tropical plants. This family, fast growing in popularity among house plant growers, includes the pineapple and many handsome ornamentals.
Bud drop. An unexplained malady which causes *gardenias, camellias* and some other plants to drop their buds.
Bulb. An underground stem, usually more or less round, from which a plant grows.
Bulbil. A small bulb borne above the ground among the leaves or flowers of a plant. When planted, it produces a new plant. (The little bulbs that develop in the ground alongside large bulbs are called bulb*lets.*)
Chopped bark. A coarse, fibrous material made from bark in which orchids are sometimes planted.
Coldest climate. For example, Minneapolis, Madison, Wis., and Bismarck, N.D.
Conifer. A tree or shrub with more or less needle-like foliage which usually bears cones and usually is evergreen. A number of trees called conifers, however, do not meet this description. For instance, yews do not have cones; bald cypress is not evergreen.
Cool climate. For example, Buffalo, Chicago, Spokane.
Corm. A form of bulb from which a plant grows.
Cormel. A small corm. If planted every year it will eventually grow into a corm.
Deciduous. Adjective describing plants that lose their leaves at some time during the year.
Dibble. A round, pointed tool for making holes in soil. It should never be used.
Divide. To separate the roots of one plant into several new plants. See How to Propagate Plants.
Division. The result of dividing plants.
Epiphytic. Growing above the ground, usually in a tree.
Evergreen. A plant that holds its leaves throughout the year. In actual fact, however, evergreens lose their leaves, but so gradually that they are never naked. Note also that evergreens are not always green; they may be purplish, bronze colored, etc. at some times of the year.
Extreme climates. The coldest and warmest climates. Everybody knows that there are many southern plants that are killed by cold. A fact that is often overlooked is that there are

also many northern plants that are killed by heat.

Eye. A bud or growing point, as in a potato tuber.

Force. See How to Force Bulbs for Indoor Bloom.

General-purpose potting soil. See How to Grow House Plants.

Greenwood. Term used to describe the young wood from which some stem cuttings are taken. See How to Propagate Plants.

Half-hard wood. Term describing the partially mature wood from which some stem cuttings are taken. See How to Propagate Plants.

Hill. A point at which seeds are grouped in the vegetable garden. Cucumbers, for instance, are often planted in hills. As a verb, hill means to mound up soil around a plant.

Humus. Decomposed vegetable matter. Probably the most important material in soil.

Insectivorous. Adjective describing plants that catch and digest insects.

Layering. See How to Propagate Plants.

Leader. The thin top of the trunk of an evergreen tree. Also the erect, uppermost branch or stem of any tree.

Leaf cutting. See How to Propagate Plants.

Leafmold. An outstandingly good form of humus.

Leggy. Adjective describing a plant that has grown tall and skinny, or gawky.

Loam. A soil composed of sand, silt and clay.

Mild climate. For example, New York City, Memphis and Denver.

Milticide. Spray or dust for killing mites.

Mulch. A blanket or covering of straw, leaves, peatmoss, etc. that is placed on the soil around plants.

Node. A joint where a leaf or bud is borne.

Nutritional spray. Fertilizer that is sprayed on foliage.

Offset. A plant or part of a plant that develops from the base of an established plant. For instance, the small bulbs that develop from, and are adjacent to, large bulbs are called offsets. The little plants that develop alongside some succulents are also called offsets.

Osmunda. A coarse, brown fiber in which orchids, bromeliads and some other plants are grown. It is sometimes called osmundine.

Peatmoss. The most widely available form of humus, sold by all garden supply stores, variety stores, supermarkets, etc. Peatmoss is very useful for lightening soils and increasing their moisture-holding capacities. All peatmoss is slightly acid, and some is very acid and sold as acid peat.

pH. A measurement showing whether a soil is acid, neutral, or alkaline. See How to Improve Soil.

Pinch. To nip off the tip ends of plant stems. In many cases, this is easily done with the fingernails, but sometimes you may have to use shears or a knife.

Plant band. A strip of heavy paper that is formed into a soil-filled square or circle in which seedling plants are grown. A cheap but good substitute for a pot.

Plunge. To sink a pot containing a plant into the ground almost to its rim. This is done to prevent the soil in the pot from drying out fast and also to prevent the pot from being overturned. When a pot is plunged, it is important to place a thick layer of gravel, sand or hard cinders directly under it so that the water can drain out of the pot and also to keep worms from crawling into the pot through the bottom drainage hole.

Potbound. Adjective describing a plant whose roots almost completely fill a pot and can hardly develop any more. Generally, when this happens, plants stop growing well. Some plants, however, actually grow better.

Pseudobulb. The swollen, or bulbous, part of the stem of an orchid.

Ray flower. A daisy-like flower; the petals growing out from the center of the flower in rays.

Rest period. The period when a plant is dormant and does not grow actively. All plants stop flowering when they rest and some lose their leaves.

Rhizome. Technically, a rootstock. As far as most gardeners are concerned, however, it is a fleshy root or type of bulb from which a plant grows.

Rootball. The roots of a plant and the soil packed around them. An established potted plant has a conical rootball. Trees and shrubs sold by nurseries usually have flattened, round rootballs.

Rootbound. Potbound (which see).

Runner. A slender stem which runs horizontally over or sometimes under the soil. It roots readily when covered with soil or where a node touches the soil.

Scale. Tiny, sucking insects with hard shells.

Scale-like. Adjective used to describe very small leaves or needles which lie close to the twig to which they are connected. Old junipers, for example, have scale-like leaves.

Seedbed. A not-too-large, outdoor area which has been spaded up and prepared for the sowing of seeds.

Semi-ripe wood. Term describing the partially mature wood from which some stem cuttings are taken. See How to Propagate Plants.

Shards. Pieces of broken flower pot.

Sidedress. To apply fertilizer next to a plant and work it into the soil.

Softwood. A term used in propagating and applied to a stem cutting which has not matured and become hard or brittle.

Sphagnum moss. A light brown, fairly fine bog moss that is sterile and moisture-holding. It is used in air-layering and as a soil medium. And when it is finely shredded, seeds may be sown in it.

Spur. A stubby, woody shoot bearing flowers and fruits. Apples grow, not directly from the limbs of the trees, but on spurs.

Starter solution. Dilute chemical given to small plants when they are transplanted to help them survive the shock of being moved. Some solutions contain special hormones; others are special fertilizer formulas.

Stem cutting. See How to Propagate Plants.

Stigma. The tip of the somewhat swollen organ centered in a flower.

Subsoil. The soil below the top, or surface, soil. It is composed mainly of inorganic matter and is not very nourishing.

Subtropics. The warmest parts of the country—southern Florida and southern California.

Sucker. An unwanted shoot rising from the roots or the lower part of the trunk of a plant. See How to Propagate Plants.

Sweet soil. Alkaline soil; any soil with a pH over 7.0.

Systemic poison. Insecticide which

is applied to the ground around plants and taken up by the roots and distributed throughout the plant. It has a relatively long-term effect.

Temperature climate. For example, Boston, New York, Detroit, Louisville, Boise, Salt Lake City, Seattle.

Terrestrial. Growing in soil.

Thin. As a verb, thin means to remove some of the branches or fruits on a plant. As an adjective, the word is often used to describe soils that are sandy, gravelly, supplying little nourishment.

Topdress. To apply some material, usually fertilizer, on top of the soil and to work it in. For instance, if you spread manure or humus on a lawn you are topdressing the lawn.

Transpiration. The loss of water to the air through plant leaves.

Tuber. A fleshy plant organ from which new growth comes.

Vermiculite. A lightweight, grey, mica-like material sometimes used as a substitute for sand, particularly in seed sowing. Another material like it is perlite.

Warm climate. For example, Atlanta, Houston, Phoenix and San Francisco.

Warmest climate. For example, Miami, Los Angeles and San Diego.

Watersprout. An unwanted shoot sprouting from the upper part of the trunk or large branches of a tree.

INDEX